TREES
TREES
TREES

OF THE EASTERN AND CENTRAL UNITED STATES AND CANADA

Dr. William M. Harlow

professor of wood technology
state university of new york
college of forestry at syracuse

Dover Publications, Inc., New York

Published in Canada by General Publishing Company, Ltd., 30 Lesmill Road, Don Mills, Toronto, Ontario.
Published in the United Kingdom by Constable and Company, Ltd., 10 Orange Street, London WC 2.

This Dover edition, first published in 1957, is a corrected republication of the work originally published in 1942 by the McGraw-Hill Book Company under the title *Trees of the Eastern United States and Canada*.

International Standard Book Number: 0-486-20395-6
Library of Congress Catalog Card Number: 57-4601

Manufactured in the United States of America
Dover Publications, Inc.
180 Varick Street
New York, N.Y. 10014

To

WILLIAM GOULD VINAL

NATURE GUIDE AND FRIEND

PREFACE

General interest in trees and forests has recently been stimulated by a new realization of their importance in erosion control and water storage, as cover for wild life, to the camper and hiker, and as a continuing source of wood and other raw materials. Many people for the first time have found that trees, their recognition and life histories, may prove to be a fascinating lifelong hobby. Few of these, however, are prepared or inclined to pursue a course in botany in order to learn the names and distinctive features of their common trees.

Most tree books are written by systematic botanists who feel that anyone who really wishes to know the trees should first learn the scientific language of the botanical fraternity. While it is true that this is the best way to train a specialist, it seems very doubtful whether the average hobbyist, hiker, camper, or woodcrafter should be expected to learn and ferret out the meaning of a large number of terms, such as polygamodioecious, zygomorphic, glaucescent, exocarp, and verticillate, before getting acquainted with the common trees. The specialist will argue that each of these terms has a definite and precise meaning which is necessary to describe properly the various plant parts. Although this is true, the author hopes to show how easy it is for the amateur to get along with the barest minimum of such terms, and how most trees can be recognized without using any of them.

Rediscovering through early writers some of the pioneer uses of our trees and their peculiar features of value to the woodcrafter has been a real pleasure. Comments or suggestions in this direction will be appreciated. Finally, wildlife uses have been included, for the most part taken from Van Dersal's compendium (see page 278), and bulletins of the Lakes States Forest Experiment Station. A large share of the illustrations have been adapted by slight reduction from those in "Textbook of Dendrology" by W. M. Harlow and E. S. Harrar.

WILLIAM M. HARLOW

CONTENTS

INTRODUCTION

What Is a Tree?—There are surely many answers. To the technically minded forester, it may be thought of as a factory for producing the most good timber in the least possible time at the least expense, a concept of the greatest importance to a wood-using nation like ours. But what are streets, parks, and playgrounds without trees or the home backyard without at least one tree to mark perhaps the change of seasons with its swelling spring buds, early summer flowers, and brilliant autumn leaves?

Trees are like children; they have a chance of living long after us. Plant an acorn where you live, and in ten years you will have become so attached to the young oak tree that you won't want to move away. A treeless world would be a sad as well as a difficult place in which to live. But once, within the memory of people still living, a great hardwood forest stretched from the Atlantic to the Great Plains, in some places so thick that one hardly could see the sun from the time when the leaves came out in spring until they fell in autumn. Then the cry was "let daylight in the swamp," and a generation of pioneers and lumberjacks felled the great trunks and often burned them to get them out of the way. That era is now past, and the forester is beginning to take over.

1

Trees are woody plants, and a woody plant is one that has a persistent stem above ground and does not wither or die back in autumn at the first heavy attack of frost. The other seed plants, with their predominantly soft tissues, are the common *herbs* of every description.

Just how large a woody plant must be to qualify as a tree is a matter of opinion. Here it is defined as reaching, when full grown, a height of at least 20 ft, with a single stem or trunk and a more or less definite crown shape. Shrubs in contrast are smaller, with several stems and no particular crown shape.

Tree Shapes.—When grown in the open, the crowns of trees tend to develop characteristic shapes: Elms are vaselike; oaks are wide-spreading; hickories often oblong; etc. At the same time, the trunks are usually short, and side branches persist nearly to the ground. Such trees make poor lumber because of the limby, hence knotty, logs obtainable. Only by competition are long clear trunks built; and at the same time, the crowns are restricted and not so typically shaped as they are in the open. At this point it should be mentioned that the maximum height and diameter for a certain kind of tree are rarely if ever from the same individual. A tree that reaches the greatest height for its kind is usually smaller in diameter than one showing the maximum size across the stump.

What Is a Species?—Since the word is so frequently used, its meaning should be made as clear as possible. The dictionary says that a species is a group of individuals agreeing in common attributes and called by the same name. For instance, all the five-needled pine trees of the northeast happen to be of the same kind and are all called *white pine*. Or a species is a group of individuals enough alike so that they might have come from the same parents. If any of you are familiar with large families, you know how much brothers or sisters may vary in looks, to say nothing of other characteristics. Therefore botanists may not agree as to the classification of certain trees. For instance, some think that black maple is a separate species of tree; others say that it is so similar to sugar maple that it should be only a *variety* of the latter.

Species are grouped together, and one of these larger groups is called a *genus* (*pl.* genera), *e.g.*, pine, oak, maple, ash. Related genera are combined to make families, and so on.

Finally, there is no such thing as "specie" connected with the names of trees. Specie is "hard cash," or coined money. The singular of the word *species* is also *species;* some amateurs and poorly trained writers on trees notwithstanding.

What Tree Makes the Best Firewood?—In camp, we usually want a quick hot blaze for such things as boiling the kettle, or a smokeless bed of coals that will hold its heat sufficiently long for broiling or like purposes. Different woods behave quite differently in this respect; but rather than learn a whole list of species together with their properties, a few principles will serve. It is assumed that the wood is all air dried to about the same moisture content; and in some of the conifers, differences in amounts of resin must be considered. But as a rule the following is true.

Woods	As Firewood Burn
1. Hard, heavy, strong	1. Slowly, leaving a bed of hot coals
2. Soft, light in weight, weak	2. Quickly, leaving only ashes

In the first group are such trees as beech, hickory, sugar maple, and white ash; in the second, poplar, basswood, spruce, and pine. Furthermore, the wood of the conifers is not suitable except for kindling (where it is ideal) because it shoots out hot sparks (especially hemlock) which may damage tents or bedding.

Tree Names.—Someone has said that trees and other plants have "nicknames" and "real names." The nicknames are known to those popularly referred to as laymen, whereas botanists and others with a more technical interest use the real (or scientific) names, which are in Latin. The principal objection to the former is the lack of universal agreement as to what should be the name of a given tree, and some trees may have from 10 to 25 different common names, depending on the locality where found. Although most people may have little use for the Latin names, they are included to protect the author in matters of species identification. A few of the longer names will be found very useful as "jawbreakers." In most cases the common and scientific names are those now recognized by the U.S. Forest Service.

How trees get their common names provides an interesting study in pioneer culture, since they were mostly named early in our history.

SOME REASONS FOR CERTAIN TREE NAMES

1. *Use:* sugar maple, paper mulberry, tanbark oak
2. *Habitat:* river birch, swamp white oak, water elm
3. *Some distinctive feature:* white birch, trembling aspen, shagbark hickory
4. *Region:* Virginia pine, Ohio buckeye, northern red oak
5. *Names from other languages*: Chinkapin oak (Indian for chestnut)
6. *Named after someone:* Douglas-fir, Engelmann spruce.

A Survey of Tree Terms

Many precise botanical terms are used to describe tree parts, but only a few are necessary to identify the common native trees.

Principal Kinds of Trees.—As we walk in the woods and look about us, it soon becomes clear that there are two great races of trees quite different from each other. First there are trees such as the pines, spruces, and hemlocks—cone bearers with needlelike or narrow leaves which stay on for more than one year, so that we call them *evergreens*—and, second, maples, birches, and others with large, spreading leaves that fall each autumn (deciduous). Woodsmen usually speak of the evergreens as *softwoods* and the broadleaved trees as *hardwoods*. Although it is true that the wood of some pines, spruce, and balsam is soft and that of oak and maple is hard, it is also true that the wood of the southern pines is harder than that of poplar or basswood (broadleaved). Moreover, the larches (cone bearers with narrow leaves) lose their leaves each fall and are bare all winter; and yew trees, although having evergreen foliage, bear a peculiar fleshy fruit instead of a cone. Finally a few of the broadleaved trees have evergreen leaves.

However, these are the exceptions, and in general we can use such terms if we remember their limitations.

Trees

Conifers	
Evergreens (persistent leaves)	Deciduous
Narrowleaved	Broadleaved
Softwoods	Hardwoods

Leaves.—After noticing whether the leaves are evergreen or deciduous,[1] the next item to observe is whether they are *simple* (one blade) or *compound* (several blades) (see Plate 1). To find out how much of a mass of foliage is a single leaf, start at

---

[1] Deciduous leaves are more or less thin, membranous, and wilt readily; persistent ones are thick and leathery, and those of the previous year are weathered or ragged.

Simple

Palmately compound

c

Pinnately compound

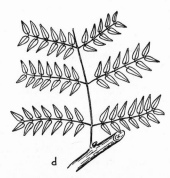

d

Twice Pinnately compound

PLATE 1.—Kinds of leaves.

the tip of what seems to be a leaf and follow it
toward the tree until you come to a *woody stem* or,
except in the late spring, a *bud*.[1] Break off everything
to this point, and you will have a complete leaf. In
a few cases, like the Kentucky coffeetree, a single
leaf is 2 to 3 ft. long with dozens of blades arranged
on a branched framework. Compound leaves with
the blades (*leaflets*) all radiating from a single point

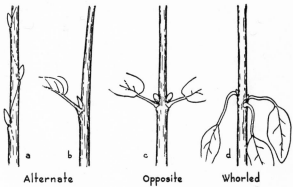

    Alternate   **Opposite**   **Whorled**

PLATE 2.—Leaf arrangement.

are called *palmate;* those arranged on both sides of a
central stem, *pinnate*. If the central stem is branched,
the leaf is then twice, or *doubly compound*, a relatively
rare thing.

Most trees have *alternate* leaves, occurring singly
in spirals (Plate 2); in a few groups, the leaves are
paired (*opposite*) or rarely in threes (*whorled*). An
easy way to remember those with opposite leaves
(with one or two exceptions) is to repeat "madcap
horse." Here the "m" stands for maple; the "a"
for ash; the "d" for dogwood (one exception); the
"cap" for Caprifoliaceae, or honeysuckle family,

[1] In a few trees the buds are submerged in the bark.

containing a few small trees and many shrubs; and the "horse" for horse chestnut (including the buckeyes).[1]

In a few alternate-leaved trees, adjacent leaves tend to become paired or staggered, a condition called *subopposite* (purple willow and buckthorn are the only ones included that show this feature). Sometimes alternate leaves may look opposite or whorled because they are on a very slow-growing branchlet or spur (see birches and cherries especially). Because of this, always look at reasonably fast-growing twigs to determine the arrangement.

Strange as it may seem, these alternate leaves are not "stuck on any which way" but are borne in almost mathematical precision. If within the same season's growth, you will pick out two leaves one of which is exactly above the other, you can prove that this is so. First, starting with the lower leaf, spiral around the twig, touching in turn each leaf stem or bud as you pass until you get to the second leaf chosen. In so doing you will probably have made one, two, or (very rare in broadleaves) three complete turns around the twig. Write this number down, and draw a line below it. Now go back; and *without* including the lower of the two leaves, count *all* the leaves passed, up to and including the second one of the pair originally chosen. Put this number below the line, and a fraction representing leaf arrangement results.

The arrangements represented by these fractions are typical of certain trees. For instance, elms are 1/2; alders 1/3 (rare type); and oaks 2/5.

The series 1/2, 1/3, 2/5, 3/8, 5/13, etc., is named for its Italian discoverer Fibonacci and shows some peculiar relationships. The added numerators of the first two terms equal the numerator of the third;

[1] A number of other shrubs also have opposite leaves, but we are here chiefly concerned with trees.

the added denominators of the first two equal that of the third; etc. There are other relationships if you look for them.

When first appearing in the spring, the leaves of certain trees are accompanied by a pair of "little leaves" (stipules) attached to the twig at the base

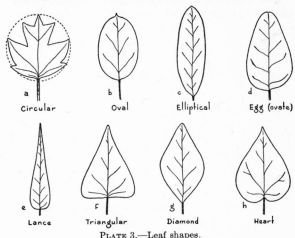

PLATE 3.—Leaf shapes.

of the leaf stem. These usually fall soon but leave scars useful in twig identification.

Leaf Shapes (Plate 3).—Some of the commonest are

1. Circular
2. Oval[1]
3. Elliptical
4. Egg-shaped (ovate)
5. Lance-shaped
6. Triangular
7. Diamond-shaped
8. Heart-shaped

Leaf Margins.—The type of margin is one of the best features for recognizing leaves (Plate 4). It

---

[1] According to the dictionary, oval may be egg-shaped; here it means between circular and elliptical.

may be (1) *smooth;* (2) *wavy;* (3) *toothed*, with sharp (serrate) or rounded (crenate), fine or coarse teeth; in the latter case, each large tooth may again bear smaller teeth; (4) *doubly toothed.* When the spaces between the teeth run one-third or more to the center of the leaf, it is then called (5) *lobed.*

Leaf Surfaces.—Sometimes these are characteristic. Most leaves are smooth and green, but some bear a whitish bloom (which can be rubbed off with the

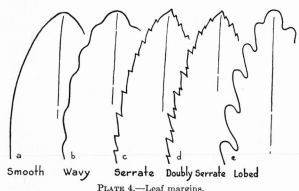

| a | b | c | d | e |
| Smooth | Wavy | Serrate | Doubly Serrate | Lobed |

PLATE 4.—Leaf margins.

finger) on the undersurface; others are hairy or velvety; and in a few, like those of slippery elm, the upper surface is rough, feeling like a piece of sandpaper.

Flowers.—Except in such trees as the tulip and magnolia, tree flowers are often inconspicuous. Who, for instance, has seen the flowers of white cedar, spruce, or elm? Moreover, flowers are available only for a short time, and whole seasons may pass without any developing at all. For these reasons, they are not of so much practical value in tree identification as are other features. However, to botanists they are of prime importance in tracing relationships between trees and tree groups; and to the miniature camera enthusiast equipped to take close-ups at an initial magnification of three to four times natural size, they offer a whole new world to explore. Many of them are not only grotesque but, in some instances, extremely colorful as well.

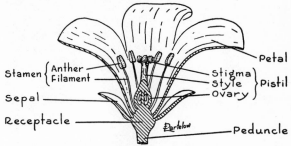

PLATE 5.—Flower structure.

The essential parts of a flower are (1) the *stamens* (pollen-bearing organs) and (2) the *pistil*(s) (receptive to the pollen and containing in its base the egg that later when fertilized becomes the seed). When both stamens and pistil(s) are found in the same flower, it is said to be *perfect* (Plate 5). But in some trees, the stamens are borne in one flower, the pistil(s) in another; in this case, the pollen-bearing flowers are called *staminate*,[1] the seed-producing *pistillate* (Fig. 1). At this point, it should be mentioned that the pistil as a structure is not found in the conifers, which belong to a great group of plants characterized by seeds that are borne naked, hence called Gymnosperms (naked seed). The protective pistil enclosing the developing seeds is characteristic of the Angiosperms, a huge aggregation of plants including our "hardwood" trees. Since there is no pistil in the

[1] Although not quite correct technically, staminate equals male, and pistillate equals female.

conifers, the seed-producing flowers cannot be called pistillate, but rather *ovulate*.

In some trees, staminate flowers and pistillate flowers are both found on the same tree. In this

Fig. 1.—Male (staminate), and female (pistillate) flowers of willow, × 15. Arrows show nectar glands which attract insects. Many of these tightly packed flowers form a catkin or "pussy."

case, any such tree may bear fruit and seeds when old enough. But some kinds of trees bear the staminate flowers on one individual and the pistillate on another. Here only certain trees can ever bear seed (the pistillate), but the presence of staminate trees is also necessary. Botanists have symbols for staminate or pistillate flowers, which we can use graphically to sum this up.

♂ represents Mars
♀ represents Venus' mirror
Ⓐ is a staminate flower
Ⓑ is a pistillate flower
Ⓒ is a perfect flower with both stamens and pistil(s)

Plate 6 shows the two principal arrangements. When separate staminate and pistillate flowers are

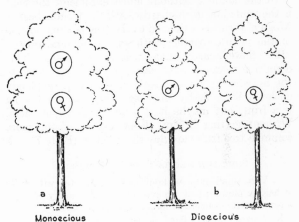

Monoecious   Dioecious
PLATE 6.—Location of flowers of separate sexes.

found on the same tree, botanists call this kind of tree *monoecious;* when on different trees, *dioecious;* and when perfect flowers *and* staminate or pistillate or both are on the same tree, *polygamous.* Perhaps this is getting too complicated; skip it if you like; none of these terms is used in the following pages.

Fruit.—This should be easy to define, but like most seemingly simple things it gets more involved the longer you work at it. The trouble is that we all think we know what a fruit should be. However, a judge has ruled that a woman selling onions on Sunday was not breaking an ordinance forbidding the sale of vegetables on that day, because an onion was not a vegetable but a fruit! If an onion is a fruit, then so is a potato (maybe you thought it was). A leading plant taxonomist (a kind of botanist), when recently cornered and asked to define the word *fruit*, finally "gave up" and reverted to the small boy's definition, "Fruits is what you eats"!

In the conifers, without much argument, the cone is the fruit; but in the hardwoods, the trouble begins. Mostly, a fruit is a ripened ovary (the bottom portion of the pistil), together with any dried remnants of other flower parts, that encloses the seeds. This definition works in most cases, but then you come to the fig where a *large number* of pistils is completely enclosed by a fleshy covering. You say that a fig is a fruit, but maybe it is a whole *cluster* of them. Without getting in any deeper, some of the more common tree fruits are described below (Plate 7).

Some Dry Fruits That Do Not Split[1]

1. Very small ones, without wings, but often with tufts of hair or plumes; look up sycamore, where many of these achenes are compounded to form a ball-shaped head.

2. Winged ones like those of maple, elm, and ash (samaras).

3. Nuts like those of hickory or oak; the covering of the nut proper is bony or leathery, usually more or less encased in an outer husk.

Some Dry Fruits That Do Split[2]

1. A flat pod (like in bean or pea) splitting on opposite sides (legume).

[1] All these have a single seed cavity in the ovary.
[2] The first two also have a single cavity in the ovary.

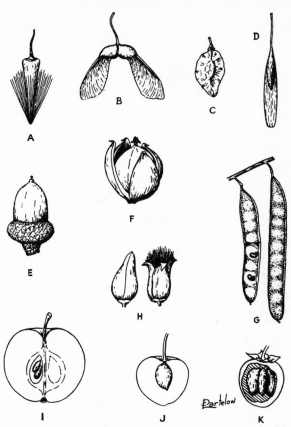

PLATE 7.—Some common tree fruits. (*A*) Small, tufted, seedlike fruit of sycamore; (*B*, *C*, *D*) winged fruits: (*B*) maple, (*C*) elm, (*D*) ash; (*E*) acorn or nut of oak; (*F*) nut of hickory; (*G*) pod of locust; (*H*) capsule of willow; (*I*) fruit of apple (pome); (*J*) fruit of cherry (drupe); (*K*) fruit of persimmon (berry).

2. Similar to 1 but splitting on only one edge (follicle). (See Magnolia, where a large number of these are compacted to form a cone.)

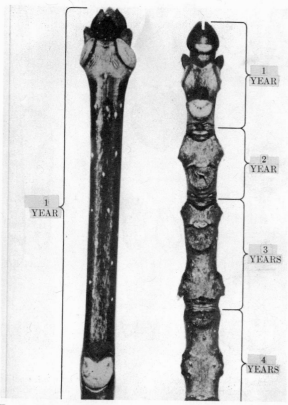

Fig. 2.—Twigs of white ash from the same tree showing differences in rate of growth.

3. A capsule like that of catalpa; derived from an ovary that may have several compartments.

SOME FLESHY FRUITS

1. Drupe (cherrylike); soft on the outside and with a bony layer enclosing the pit or seed.

2. Pome (applelike); fleshy without, the seeds protected by a papery layer; you have all had portions of this in your mouth when eating an apple too close to the core, or in apple pies.

3. Berry; soft and fleshy with the seeds scattered throughout. Blueberry, tomato, and persimmon are examples, but blackberry, raspberry, and mulberry are *not*. The last three are made up of miniature druplets collected together.

In many cases, it is not possible to tell from inspection just what kind of fruit you have, and a careful study is necessary of the flower from which it was derived. This is something that you must take a course in botany to appreciate.

Twigs.—Twigs together with their buds can be used for identifying trees during 8 to 9 months of the year, since next season's buds are formed in July, a fact quite new to most people who think that the buds suddenly pop out from the twigs in early spring. To get very far with twig study, a pocket magnifying lens is necessary. This field is outside the scope of the book,[1] but some features are visible to the unaided eye and are therefore included (Plate 8). Incidentally, students seem to think that twig features are easier to learn and use than almost any others.

Bud Arrangement.—Just as with leaves, the arrangement, whether opposite, whorled, alternate, or subopposite, is the first thing to notice.

Terminal and Lateral Buds.—Some trees develop on the ends of their twigs true terminal buds; others

---

[1] See W. M. Harlow, *Twig Key to the Deciduous Woody Plants of Eastern North America.* Published privately.

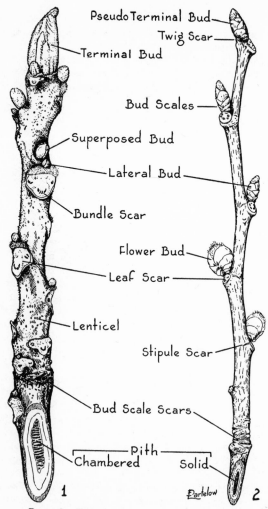

Pseudo Terminal Bud
Twig Scar
Terminal Bud
Bud Scales
Superposed Bud
Lateral Bud
Bundle Scar
Flower Bud
Leaf Scar
Lenticel
Stipule Scar
Bud Scale Scars
Pith
Chambered   Solid

Bartelow

1   2

PLATE 8.—Twig details of (1) butternut, (2) elm.

do not. A true terminal bud usually points straight upward and is commonly larger than the side buds (laterals). In those trees where no terminal is produced, one of the side buds occupies the end position; since it is really a side bud, it rarely points straight ahead but rather slightly to one side and is about the same size as those below it. However, the only sure way to know that it is a false or pseudo-terminal is to find a small round scar at its base (not a leaf scar, see below) where the advance growth has died back and fallen off. Basswood and sycamore are good examples, and here these branch scars can be seen without a lens.

Fig. 3.—Basswood bud showing branch scar × 4.

Bud Coverings.—Buds are in reality miniature branchlets with tiny leaves or flowers already formed and waiting to fulfill the prophecy "if winter comes, can spring be far behind?" In some way or other, these delicate structures must be protected not so much from cold as from drying out. This is accomplished in several ways. In a few cases, the first pair of leaves is thick and waterproof and encloses the inner ones completely; this kind of bud is called *naked*. In most trees, however, bud scales have been developed for protection. They vary in number from one to many, and this number is usually typical for the kind of tree in question. For instance, willows have a single scale which in spring "unzippers" behind and is lifted off like a tiny hood; soft maples have several outer scales; and oaks, many of them arranged in five rows. Some buds are characterized by having the outer pair of scales just meet in a vertical line without overlapping (called *valvate);* but in most species, the scales overlap (like the petals of a rose).

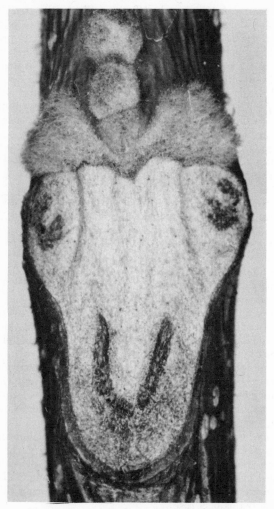

Fig. 4.—Butternut leaf scar showing bundle scars.

Leaf Scars.—Long before leaf fall, a special layer that produces cork develops across the base of each leaf stem where it joins the woody twig. When autumn comes and the leaves are driven away before the "west wind" "like ghosts from an enchanter fleeing," the exposed spots that they leave are protected from losing moisture. This usually small *leaf scar* varies greatly in size and also in shape and bears on its surface the sealed

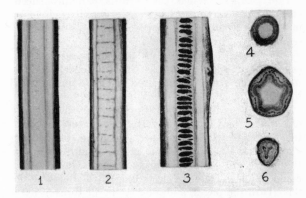

Fig. 5.—Pith types and shapes. *1.* Solid; *2.* solid with cross bars (diaphragmed); *3.* with cross partitions (chambered); *4.* circular; *5.* angled or star-shaped; *6.* triangular.

ends of the tiny tubes through which the sap passed from the twigs to the leaves, and the leaf-made food flowed back to the growing parts of the tree. These tiny calloused ends, looking something like pinheads, are called *bundle scars.* Their number, size, and arrangement are often useful in identifying the twigs that bear them. In a few cases, they are prominent to the naked eye, but usually a lens is needed to appreciate them. Magnified leaf scars are often very grotesque objects, especially those like the one of the butternut (Fig. 4).

Pith Types.—Most twigs, when whittled length-wise, show a solid pith varying in color with the kind of tree chosen. In a few, however, such as those of tulip and black tupelo, indistinct bars of darker tissue cross at intervals; this type is called *dia-phragmed.* Finally, a few species have a hollow pith interrupted by numerous cross partitions. The walnuts are characterized by this *chambered* type.

Pith Shapes.—When the pith is carefully sliced crosswise, its shape can be seen. In elm, it is circular; in alder, triangular; and in poplar, oak, and hickory, star-shaped or five-angled.

Twig Color.—This may be an important feature, especially in such groups as the dogwoods. Colors may include green, purple, red, yellow, orange, and brown.

Bark.—This is probably the most variable feature of all, and the Chinese proverb "one picture is worth ten thousand words" is especially applicable. Bark varies particularly with tree age, but also with site (habitat) and geographical location. Many terms could be listed here for describing bark, but why do it? Turn the pages, and look at the pictures of bark that follow—and don't forget that many trees that you will see won't look exactly like the illustrations given! If space permitted, there should be from two to six or more bark pictures for each tree. After ten years or more in the same region, you may get so that you never make a mistake on bark features; but this is what keeps it interesting.

How to Make Leaf Prints

Leaf prints are easily made using only some printer's ink (obtainable in most printer's shops, often for the asking) and a piece of cardboard.

Rub a small amount of the ink into the cardboard until a smooth, inked surface is produced without ridges or patches of excess ink. Place a leaf, under-side down, upon the inked cardboard; cover it with a piece of newspaper; and carefully rub, don't pat, for several seconds until you have been over all of the leaf surface.

Then remove the paper; fold it with the inky side in; and throw it in the wastebasket! Printer's ink has a great tendency to fasten itself to the fingers

Fig. 6.—Four steps in making leaf prints.

if given half a chance. You want leaf prints, not fingerprints.

Now carefully remove the inked leaf, and place it upon a piece of typewriter or similar paper. Cover it with a clean piece of newspaper, and repeat the rubbing process, but be careful that meanwhile the leaf does not slip, because this will cause a blurred image. Now lift off the paper; pick up the leaf; and see what you have. A little practice will ensure the best results. Too much ink on the board results in black patches on the print; too little makes the

prints too faint. You can also use printer's ink rolled
out with a rubber roller upon a sheet of glass.

How to Use the Keys

A key, in this case, is a series of signboards or
clues that lead you, sometimes through a forest of
unknowns, to a particular tree or group of trees.
Following such a trail develops that keenness for
observing detail which is the mark of the true detec-
tive. When you discover a new tree whose identity
you would like to learn, the keys should enable you
to do so. However, it is of the greatest importance
that you select typical material. In most cases, the
keys are based on leaf features, but flowers, fruit,
twigs, and general appearance may also be included,
and this means that you should observe all that you
can of these features as well. To get started, select
a leaf or, in some of the conifers, a branchlet that
appears to be "average." Avoid unusually large or
small ones, and especially do not choose stump
sprouts, since these often bear leaves quite different in
size, shape, and arrangement from average growth.

We now approach the beginning of the general
key (see below). Notice that we must go in one of two
directions, no straddling possible. Let us assume
that the leaf chosen is deciduous, opposite, simple,
and cut into lobes that radiate from the base (palmate). Since it is not evergreen, the second No. 1
directs us to No. 8. Here it is clear that the leaves
are not at all needlelike, so the second No. 8 sends
us to No. 9. The leaves being opposite, the search
is directed from the first No. 9 to No. 10. The leaves
are simple; so No. 11 is indicated. Here the first part
reads "Leaves lobed palmately like fingers on a
hand"; and since this describes our unknown leaf,
we know that it is some kind of maple. Turning

now to the species key to maples, we can find out what kind of maple it is. In this case, more specific features would be needed to complete the identification. As a final check, compare the leaf with the illustration and description given in the book. No matter how carefully made, keys are never foolproof until used by many people and in different localities. If after careful use there seem to be "rough spots," suggestions will be welcome, accompanied with samples of the material in question.

GENERAL KEY TO TREE GROUPS

(Before using this key *make certain* that you do not have poison sumac, poison ivy, or poison oak (p. 230); these are not included in the key.)

1. Leaves, needles, or foliage, evergreen[1] **2**
1. Leaves or foliage, deciduous (falling at the end of the growing season) **8**

    2. Leaves, ½″ or more in width, broad . . . . **Holly** (p. 235), **Sweetbay** (p. 185), **Rhododendron and Mountain-laurel** (p. 263)
    2. Leaves, less than ½″ wide, needlelike, narrow, or small and scalelike **3**

3. Leaves, small and scalelike, close together, and overlapping . . . . **Cedars** (p. 69)
3. Leaves, not scalelike, long and narrow **4**

    4. Leaves, needlelike, in 2s, 3s, or 5s, united at the base to form bundles; when held together, the needles of each bundle form a cylinder . . . . **Pines** (p. 33)
    4. Leaves or needles, not in bundles, but occur singly **5**

---

[1] In summer look for both old, leathery leaves or needles, and new fresh ones on the same twig.

5. Needles, paired or in 3s around the twig..........
   Eastern redcedar and Oldfield **juniper** (p. 74)
5. Needles or leaves, alternate, in spirals, not opposite
   each other.................................... **6**

    6. Leaves, needlelike, 4-sided (in one species
       diamond-shaped) in cross section (roll between
       thumb and finger to feel edges.**Spruces** (p. 55)
    6. Leaves, blunt, with essentially parallel sides,
       flat in cross section......................... **7**

7. Twigs, stout; buds, sticky; older twigs show small
   circular scars where leaves have fallen off; top of tree,
   stiff and spirelike...............**Balsam fir** (p. 66)
7. Twigs, slender; buds, not sticky; top of tree, flexible,
   bends over.....................**Hemlock** (p. 63)

    8. Leaves, narrow, almost needlelike, many times
       longer than wide, on old twigs occur in tufts on
       woody spurs...............**Tamarack** (p. 52)
    8. Leaves, broader, not at all needlelike.......... **9**

9. Leaves, opposite or in 3s........................ **10**
9. Leaves, alternate (look on normally fast-growing
   twigs; on dwarfed growth, the crowded leaves may
   appear opposite when really alternate)............ **17**

    10. Leaves, simple............................. **11**
    10. Leaves, compound.......................... **16**

11. Leaves, lobed palmately like fingers on a hand.....
                                 Maples (p. 236)
11. Leaves, not lobed............................. **12**

    12. Leaf margin, smooth, not toothed............. **13**
    12. Leaf margin, toothed (serrate)............... **15**

13. Side veins parallel margin.......**Dogwoods** (p. 259)
13. Side veins do not parallel margin................. **14**

14. Leaves, heart-shaped, usually in whorls of 3....
Catalpa (p. 274)
14. Leaves, elliptical, paired....Fringetree (p. 266)

15. Side veins parallel margin......Buckthorn (p. 253)
15. Side veins do not parallel margin. Viburnums (p. 276)

16. Leaves, palmately compound (see No. 11).....
Buckeyes (p. 250)
16. Leaves, pinnately compound..Boxelder (p. 246)
Ashes (p. 266)

17. Leaves, simple.................................. 18
17. Leaves, compound............................... 50

18. Leaves, lobed............. 19
18. Leaves, unlobed.......... 26

19. Undersurface of leaf, covered with silvery wool....
White poplar (p. 97)
19. Undersurface, not silvery-woolly................. 20

20. Ends of lobes each bear a bristle or hair tip....
Red oaks (p. 141)
20. End of lobes, not bristle-tipped............... 21

21. Leaves and twigs have a spicy odor and flavor.....
Sassafras (p. 189)
21. Leaves and twigs, not spicy.................... 22

22. Outline of leaf, elliptical or broadest above the
middle..................White oaks (p. 141)
22. Outline of leaf, circular or nearly so........... 23

23. Base of leaf stem, hollow, enclosing the next year's
bud........................Sycamores (p. 195)
23. Base of leaf stem, solid....................... 24

24. Sap of broken leaves and twigs, milky (cloudy)
Mulberries (p. 178)
24. Sap, clear, not cloudy....................... 25

25. Leaves, mostly 4-lobed, the apex "chopped off" or indented with a wide notch......**Tuliptree** (p. 185)
25. Leaves, 5- or 7-lobed, star-shaped.**Sweetgum** (p. 191)

   26. Leaf stem, flattened so that the leaf trembles in the slightest breeze.....**Poplars, aspens** (p. 85)
   26. Leaf stem, circular or grooved in cross section.. **27**

27. Leaf margin, entire, not toothed in any way....... **28**
27. Leaf margin, serrate or with rounded teeth........ **33**

   28. Leaves, tipped with a bristle or hair...........
           Shingle and willow oaks (pp. 165, 167)
   28. Leaves, lacking a bristle at the end........... **29**

29. Sap, milky (break leaf stem); twigs, armed with sharp thorns................**Osageorange** (p. 181)
29. Sap, clear; twigs, unarmed..................... **30**

   30. Leaves and twigs, with a spicy odor and flavor
                  Sassafras (p. 189)
   30. Leaves and twigs, not spicy.................. **31**

31. Leaves, heart- or kidney-shaped...**Redbud** (p. 217)
31. Leaves, elliptical to oval or widest above the middle **32**

   32. Pith, when sectioned lengthwise, shows faint cross bands of darker tissue....**Black tupelo** (p. 258)
   32. Pith, either chambered or, if solid, without cross bands......**Cucumbertree** (p. 182), **Pawpaw** (p. 188), **Persimmon** (p. 264), **Alternate-leaved dogwood** (p. 263)

33. Leaf margin, with coarse wavy or rounded "teeth"
    Swamp white oak (p. 151), **Chestnut oak** (p. 150), **Witchhazel** (p. 194)
33. Leaf margin, with sharp teeth; or if these are rounded, they are very small.................... **34**

34. Sap, milky or cloudy......**Mulberries (p. 178)**
34. Sap, clear.................................. **35**
   (From here to No. 50 the trees get "taller and
   closer together" and the trail is dimmer; only
   the alert will not get lost!)

35. Leaves, average 4″ or more in diameter, nearly
   circular, somewhat heart-shaped at the base......
   Basswood (p. 255)
35. Leaves, not circular or, if so, less than 4″ in diameter **36**

36. Leaves, with medium-sized to large definitely
   single teeth on margin....................... **37**
36. Leaves, with double teeth or such small ones
   that it is difficult to see whether they are single
   or double................................... **40**

37. Leaves, lopsided and more or less heart-shaped at
   the base; fruit, a large-pitted drupe, the thin flesh
   tasting like a date............**Hackberry (p. 176)**
37. Leaves, more or less equal at the base, not heart-
   shaped; fruit, a nut.......................... **38**

38. Teeth, ending in a hair or bristle.............
   Chestnuts (p. 136)
38. Teeth, without bristles...................... **39**

39. Teeth, sharp; leaf, with a papery rattle; buds, long
   and lance-shaped.................**Beech (p. 133)**
39. Teeth, slightly rounded like a nipple; buds, short
   and egg-shaped........... **Chinkapin oak (p. 153)**

40. Twigs, armed with long thorns....**Thornapple
   (p. 206)**, **Wild pear (p. 209)**, **Flowering crab
   (p. 208)**
40. Twigs, without thorns or spines.............. **41**

41. Leaves, with conspicuous medium-sized to large,
   double teeth......................**Elms (p. 168)**
41. Leaves, with *small* single or double teeth.......... **42**

42. Twigs, stout; (see illustration, 92) pith, 5-angled in cross section; leaves, broadly egg-shaped to almost circular..................................
   Balsam poplar (p. 93), **Balm-of-Gilead** (p. 95), **Swamp Cottonwood** (p. 95)

42. Twigs, slender; pith, triangular, circular, or so small as not to be easily seen; leaves, narrowly egg-shaped, lance-shaped, or elliptical to oval.. **43**

43. Pith, conspicuously triangular when sliced crosswise (make several sections)...........**Alders** (p. 129)

43. Pith, circular or so small as not to be easily seen... **44**

44. Twigs, with an intensely bitter quinine taste or bitter-almond flavor........................ **45**

44. Twigs, not as above........................... **46**

45. Twigs, bitter; bud, covered by a single scale; seeds, very small, silky haired............**Willows** (p. 77)

45. Twigs, with a faint to strong bitter-almond flavor; bud, with 2 or more scales; seeds, not silky haired, enclosed in a fleshy fruit.......................
   Cherries, Plums, Peach, Shadbush (pp. 199 to 206)

46. Veins parallel margin; fruit, fleshy............
   Buckthorn (p. 253)

46. Veins do not parallel margin; fruit, dry or fleshy **47**

47. Twigs have a peculiar sweetish taste (not wintergreen); leaf, crinkled, more or less hairy; fruit, an apple........................**Wild apple** (p. 207)

47. Tree, without the preceding combination.......... **48**

48. Bark, papery; short, dwarfed twigs (spur shoots) bearing tufted leaves common on older growth; fruit, a very small, 2-winged nutlet ("seed") borne in a cone...............**Birches** (p. 116)

48. Bark, smooth and blue-gray or finely shreddy; spur shoots, lacking; seed, unwinged.......... **49**

49. Bark, smooth and blue-gray; trunk, "muscular" appearing; nutlet, backed by a 3-lobed leafy bract..
American Hornbeam (p. 132)

49. Bark, shreddy; nutlet, enclosed in a papery envelope
Hophornbeam (p. 130)

50. Leaves, twice or thrice compound............ **51**
50. Leaves, once compound...................... **53**

51. Leaves, large, 1–3 ft. long...................... **52**
51. Leaves, smaller, less than 1 ft. in length; twigs, with long, sharp, usually 2- to 3-branched thorns.......
Honeylocust (p. 214)

52. Leaflets, entire; fruit, a thick short pod........
Coffeetree (p. 219)

52. Leaflets, serrate; fruit, a small drupe; the stout twigs with numerous short sharp spines.......
Devil's walking stick (p. 257)

53. Leaves or twigs, when broken, exude a milky sap...
Sumacs (p. 228)

53. Sap, clear.................................... **54**

54. Leaflets, 3 in number; when crushed, with a rank somewhat orange-peel odor....**Hoptree** (p. 225)
54. Leaflets, 5 or more.......................... **55**

55. Twigs, armed with spines or thorns.............. **56**
55. Twigs, unarmed.............................. **58**

56. Twigs, with long, branched thorns; leaflet margins, finely toothed; fruit, a mahogany-colored, twisted pod..............**Honeylocust** (p. 214)
56. Without the preceding combination; paired spines usually present...................... **57**

57. Crushed leaves, with a strong orange odor; small spines, on the leaf stem........**Pricklyash** (p. 223)
57. Leaves, without orange odor; spines, on woody twig only............................**Locusts** (p. 220)

58. Crushed leaves, with a peculiar, disagreeable odor like popcorn with rancid butter or, as a student once said, "like a zoo"; leaflets, toothed only near the base..... **Tree-of-Heaven** (p. 226)
58. Leaves, fragrant or odorless; margins, entire or toothed all the way along.................... **59**

59. Leaflets, entire; base of leaf stem, hollow, enclosing next year's bud; fruit, a beanlike pod.............
 Yellow-wood (p. 223)
59. Leaflets, serrate; buds, visible; fruit, a nut, or small red "apple"................................... **60**

60. Second year's pith shows chambers when sliced lengthwise.................. **Walnuts** (p. 97)
60. Pith, solid................................. **61**

61. Leaves, more or less fragrant when crushed; leaflets, large, mostly 3″ or more in length; fruit, a nut.....
 Hickories (p. 103)
61. Leaves, not fragrant; leaflets, small, mostly about 2″ long; fruit, a small red apple **Mountain-ashes** (p. 211)

THE CONIFERS OR SOFTWOODS

THE PINE FAMILY

THE PINES

Leaves (needles) are persistent for two or more seasons; in bundles of two to five, when first borne in the spring each bundle enclosed at the base in a somewhat scaly sheath.[1] Since the bundle is circular in cross section, the shape of each needle is determined by the number of needles in a bundle (see pages 36 and 41).

Flowers appear as small cones, male and female on the same tree; the former are often so abundant that at shedding time if one shakes a branch, he becomes enveloped in a yellow mist of countless billions of pollen grains and near-by ponds and lakes are streaked with them. Since pines have "on" and "off" years for flowering, several seasons may pass before such abundance is observed on any particular tree.

Fruit is a pendent or spreading woody cone which requires two growing seasons to mature. Small at the end of the first season and upright at the branch tip, it slowly turns downward the second spring, grows rapidly, and attains full size by the end of the summer. The presence of small cones means that a seed crop may be expected the next year; their absence shows that there will be no mature cones

[1] In the *soft pines*, this sheath falls off during the summer or autumn, but the *hard pines* always retain theirs. This difference is useful in separating the two groups.

at that time. If one is gathering seed, this way of predicting a year ahead is very useful.

Remarks.—The pines are among our most important forest trees, producing as they do not only wood of the finest quality but also (certain species) naval stores (turpentine and rosin). They are most common and often form extensive forests on sandy soils, with a minimum of moisture. On heavier soils they occur mixed with the broadleaved trees.

Fig. 7.—Three stages in the development of a pine seedling, showing seed leaves and first juvenile leaves. The bundled needles appear later. (*Photograph by B. O. Longyear.*)

Pine wood makes good kindling, and especially the resinous pine knots obtained from old stumps or rotting logs will ensure a fire on a rainy day. These knots can also be used as torches and will burn for a considerable time. The quick hot fire produced by pine wood leaves only ashes rather than live coals and blackens the cooking utensils with soot and tar. Like other conifers or softwoods, hot sparks may shoot out from the burning sticks, a feature not desirable in a tepee or near other tents.

Pine seeds are eaten by squirrels and several species of birds, and the young branches are browsed by deer. In winter, rabbits often attack young trees

in plantations and sometimes trim them completely
of needles or even nibble off the branches as well.

Key to the Pines

1. Needles, in 5s, on second-year growth lacking a sheath
   at the base of the bundle........**White pine** (p. 35)
1. Needles, in 2s or 3s; sheath present on all bundles.... **2**

   2. Needles, in 2s................................. **3**
   2. Needles, in 3s................................. **5**

3. Needles about 5″ long, snapping cleanly when bent
   double.....**Red pine** (see also shortleaf pine) (p. 40)
3. Needles, mostly less than 3″ long................. **4**

   4. Needles, sharp-pointed, mostly blue-green; bark,
   bright orange.................**Scots pine** (p. 51)
   4. Needles, sharp or dull, yellow-green; bark, grayish
   or dull.......**Table Mountain pine** (p. 50), (cones
   with a large spike on each scale; Pa. southward),
   Jack pine (p. 44) (minute prickle), **Scrub pine**
   (p. 49) (sharp prickle), (ranges of last 2 do not
   overlap)

5. Needles, stiff and twisted, often in tufts along the
   trunk; cones persistent for many years......**Pitch
   pine** (p. 42)
5. Needles, flexible and straight (many also in 2s on the
   same tree)....................**Shortleaf pine** (p. 47)

Eastern White Pine

(*Pinus strobus* L.)

Appearance.—A tall forest tree (largest of eastern
conifers), from 80 to 100 ft. in height and 2 to 3½ ft.
in diameter (max. 220 by nearly 6 ft.).[1] Each year's
growth is marked by a new false whorl of side branches.

[1] According to Sargent, near Merrimack in 1736, a white
pine was cut measuring 7 ft. 8 in. in diameter at the butt.

For descriptive legend see opposite page.

Especially in dense woods, old trees have straight clear trunks bearing crowns of graceful plumelike branches.

Needles.—In 5s (the only five-needled pine in the east), 3 to 5 in. long, flexible, dark blue-green; bundle sheath falling off after the first season.

Flowers.—Male and female occur separately as small cones on the same tree.

Cones require two seasons to mature, at the end of the first season about $\frac{3}{4}$ in. long, upright. They become pendent at the beginning of the second season and grow to a mature length of 4 to 8 in.; prominently stalked; scales, thin, unarmed, each bearing two terminally winged seeds.

Bark.—At first dark green, smooth, and thin; later breaks up into wide plates; on old trees, deeply furrowed into blocky, rectangular ridges.

Habitat.—Although making best growth on sandy loam soils, this tree is found on a wide variety of sites. Originally, it was found in pure stands as well as in mixture with other species.

Distribution.—The Lake states, southern Canada, the northeast, and along the Appalachians to northern Georgia.

Remarks.—Since colonial times, white pine has been the most important conifer of the northeast and Lake states. Occurring in large quantities and possessing a soft, durable, easily worked wood, it was the favorite timber for construction of all sorts.

Natural reproduction is good. Also millions of trees have been planted for reforestation purposes. When strips of land are left along the edges of plantations, fruit-bearing shrubs seed in and make an

---

Fig. 8.—Eastern white pine. *1.* First year's cone $\times \frac{3}{4}$. *2.* Mature cones and foliage $\times \frac{1}{3}$. *3.* Opened cone $\times \frac{3}{4}$. *4.* Seed $\times 1$. *5.* Bundle of needles $\times \frac{3}{4}$. *6.* Cross section of needle $\times 35$. *7.* Bark of young tree. *8.* Bark of old tree.

attractive place for wild life. White pine, especially in pure stands, is often greatly damaged by the pine weevil whose larvae riddle and destroy the terminal leader. One or more of the side branches then endeavor to take the place of the dead leader, and in this way a crooked stem results. White pine blister rust also causes damage and often death, particularly in young trees. This is a fungous disease brought over from Europe and now firmly established here. One stage is found on the leaves of gooseberry and currant bushes, and from them spores can infect pine trees to a distance of about 1,000 ft. The stage found on the pine produces a bark canker that eventually kills enough bark to cause the death of the tree. From pustules on the canker, spores may travel many miles to reinfect currant or gooseberry leaves. These spores cannot injure white pine; therefore by eliminating gooseberry and currant bushes from white pine areas the disease can be controlled.

According to Josselyn, an early English writer, "the distilled water of the green cones taketh away wrinkles in the face, being laid on with cloths." Can it be that our modern beauty experts have overlooked something?

In the colonies, after 1691, a fine was imposed for cutting white pine for mast timbers on any but private land. Such mast timber was reserved for the crown, and sometime later (1719) it became the practice (Maine) for the king's surveyors to mark these trees with an "R" (royal) or a broad arrow.

In Pennsylvania, especially from 1840 to 1860, special rafts of great mast timbers, each 90 ft. long and often 40 in. in diameter, were floated down the Susquehanna River through rapids where the water in some places drops 400 ft. in a mile. Since the mast spars must have no holes in them, they were bound together with hickory withes, instead of being pinned as were squared timbers. Such spar rafts

Fig. 9.—Old-growth eastern white pine.

were steered with mighty sweeps (oars) fore and aft and skippered by men who knew every whirlpool

and cross current for 200 miles of river. The spars
in these rafts would weave back and forth in rough
water so that the raft at a distance looked like a
piece of fabric hurtling through the rapids, and it
took a man with "educated feet" to stay aboard.
Stopping such a juggernaut for the night is a whole
story in itself.[1]

When land was cleared, the old stumps and attached
roots were set up on edge with roots interlaced, and
these fences stood for many years.

White pine inner bark in May and June is good to
chew, and New Englanders used to candy strips of it.

Red Pine   Norway Pine

(*Pinus resinosa* Ait.)

Appearance.—A tall forest tree, from 50 to 80 ft.
in height and 2 to 3 ft. in diameter (max. 120 by 5 ft.).
The crown is somewhat open, oval in outline and is
supported by a long, well-formed cylindrical bole
which is usually clear of branches. Each year's
growth is marked by a new false whorl of branches,
a feature also found in white pine; in the other two
northeastern hard pines, from one to three whorls
may be produced each year.

Needles.—In 2s (rarely also in 3s on shoots infested
with the larvae of the pine tip moth), 4 to 6 in. long,
flexible, dark yellow-green, straight, snapping cleanly
when bent double between thumb and fingers;
bundle sheath persistent.

Flowers.—Male and female are borne separately
as small cones on the same tree.

Cones require two seasons to mature. During
the first season they are small and upright. The
second season, they become pendent and, when
mature, are ovoid and measure from 1½ to 2¼ in.

[1] J. D. Tonkin. *The Last Raft.* Harrisburg, Pa. 1940.

in length. Scales, thin and unarmed, each bearing two terminally winged seeds.

Bark.—On young trees flaky and orange-red; eventually breaks up into flat plates scaly on the surface.

Habitat.—Found mostly on light, sandy loam soil of a poorer quality than is necessary for white pine.

Fig. 10.—Red pine. *1.* Bundle of needles × ¾. *2.* Cross section of needle × 35. *3.* Open cone × ¾. *4.* Closed cone × ½. *5.* Seed × 1. *6.* Bark of old tree.

Distribution.—The Lake states, southern Canada, and the northeast as far south as northern Pennsylvania (also an isolated patch in northeastern West Virginia).

Remarks.—How this native North American tree happened to have the name *Norway* applied to it seems to be somewhat of a mystery. Some say that the early voyageurs on the Great Lakes mistook it for Norway spruce, a European tree. Others

claim that this species was so named because of large stands near the village of Norway, Me. At Hancock, N. H., is Norway Pond. The author was much interested to find a number of large "Norway pines" on its shores. In any event, the name Norway pine has been used so long that the decision of the U.S. Forest Service to use the more logical name *red pine* met with considerable criticism. This tree is now widely planted on reforested areas and is a valuable timber species. According to Gibson, the heartwood (similar to that of white pine but harder) was used for ships' decks, and the trunks for masts.

Fig. 11.—Old-growth red pine left after logging.

Pitch Pine

(*Pinus rigida* Mill.)

Appearance.—Quite variable as between different parts of its range. Probably reaches best development in Pennsylvania where it is often a forest tree 50 to 60 ft. high and 1 to 2 ft. in diameter (max. 100 by 3 ft.). Through New England pitch pine is more often a small scraggly tree. Unlike red pine, more than one whorl of side branches may be produced each year.

Needles.—In 3s, 3 to 5 in. long, yellow-green, rather stiff; often twisted; bundle sheath, persistent.

Fig. 12.—Pitch pine. *1.* Bundle of needles × ¾. *2.* Cross section of needle × 35. *3.* Open and closed cones, respectively × ¾. *4.* Seed × 1. *5.* Bark of old tree.

Flowers.—Male and female occur separately as small cones on the same tree.

Cones require two seasons to mature (for details of growth see description of red pine), 2 to 3½ in. long, somewhat ovoid, usually persistent on the tree for many years after maturity (a good feature for identifying the tree at a distance); each scale armed with a sharp prickle and bearing two terminally winged seeds.

Bark.—At first dark and scaly or ragged; later breaks up into brownish-yellow plates separated by narrow fissures. The trunk often bears tufts of needles on short "water sprouts," a very characteristic feature.

Habitat.—Found most typically on poor, dry, sandy soils too sterile for most other trees except scrub oak (*Quercus ilicifolia*) and gray birch, which are its most common associates.

Distribution.—From Maine (few areas in southern New Brunswick) and the islands in the upper St. Lawrence River, south to northern Georgia and Virginia.

Remarks.—The cones open at irregular intervals, some during the winter, and the seeds are cast upon the snow, thus providing food for birds and small mammals such as the red squirrel, for which it is said to be the preferred food. Young trees when cut down or damaged by fire often produce sprouts, a feature that is rare among conifers and especially pines. Before the Revolution and the opening of the southern pineries, tar and turpentine were made from pitch pine.

According to Emerson the wood is durable even when subjected to alternate wetting and drying; for this reason, it was used in making water wheels for the gristmills so common in colonial times. The wood makes a hot fire (but sooty, of course) on account of the resins in it, and was used widely in firing steam engines.

Pitch pine will grow near the coast where occasional high tides even cover the roots.

Jack Pine

(*Pinus banksiana* Lamb.)

Appearance.—Usually occurs as a small to medium-sized tree of ragged outline, but at its best 70 to 80 ft. high and 12 to 15 in. in diameter (max. 90 by

2 ft.). Like pitch pine, this species produces two or three false whorls of side branches each year.

FIG. 13.—Jack pine. *1.* Male flowers shedding pollen × 1. *2.* Closed cones and foliage × ½. *3.* Open cone × ¾. *4.* Seed × 1. *5.* Bundle of needles × ¾. *6.* Cross section of needle × 35.

Needles.—In 2s, ¾ to 1½ in. long, yellow-green, divergent, stout, often twisted; bundle sheath, persistent.

Flowers.—Male and female occur separately as small cones on the same tree.

Fig. 14:—Open-grown Jack pine.

Cones require two seasons to mature (for details of growth see description of red pine), 1½ to 2 in.

long, oblong-conical, usually pointing forward, and often curved toward the twig, persistent for many years; each scale armed with a very small prickle and bearing two terminally winged seeds.

Bark.—Scaly, dark gray to reddish brown.

Habitat.—The poorest of dry sandy soils, in this respect resembling pitch pine. On better sites this species occurs mixed with red pine or is replaced by it.

Distribution.—Mostly a Canadian species extending from Quebec and Ontario northwest to central Mackenzie; in the United States, found in Maine, northern New Hampshire, northern Vermont, northeastern New York, and the Lake states.

Remarks.—Like the western lodgepole pine which it resembles, this species holds its cones, many of them unopened, for a number of years. Often, these unopened cones become buried in the wood as the tree grows and disappear from sight. Seeds from such cones have been found fertile after extraction. A light fire serves to open the persistent closed cones and results in pure stands of Jack pine in many localities. Some of the early settlers thought that this pine was a witch tree and that it was dangerous to get nearer than 10 ft. from it. They also thought that it poisoned the soil—perhaps because it grew naturally on poor sites.

Shortleaf Pine

(*Pinus echinata* Mill.)

Appearance.—A forest tree, at its best from 80 to 100 ft. high and 2 to 3 ft. in diameter (max. 130 by 4 ft.). The clear, well-formed bole supports a more or less pyramidal crown of small branches. More than one whorl of laterals is produced each year.

Needles.—Commonly in 2s (also 3s on the same tree), 3 to 5 in. long, flexible, dark yellow-green; bundle sheath, persistent.

Flowers.—Male and female occur separately as small cones on the same tree.

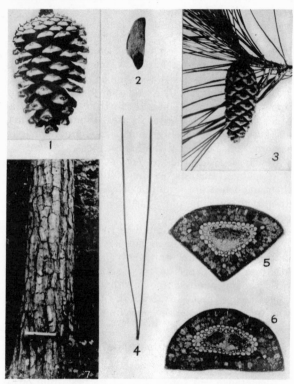

Fig. 15.—Shortleaf pine. *1.* Open cone × ¾. *2.* Seed × 1. *3.* Closed cone and foliage × ½. *4.* Bundle of needles × ¾. *5* and *6.* Cross sections of needles × 35. *7.* Bark of old tree.

Cones require two seasons to mature (for details of growth see description of red pine), 1½ to 2½ in. long, oblong to narrowly ovoid, each scale armed with a small but sharp prickle and bearing two terminally winged seeds.

Bark.—At first nearly black, roughly scaly with small surface pockets or holes, later reddish brown and broken into irregular flat plates.

Habitat.—For the most part found on dry upland soils, in pure stands or mixed with hardwoods, especially some of the oaks.

Distribution.—Staten Island (New York) through New Jersey and Pennsylvania to southern Ohio, south to eastern Texas and south central Georgia (sparse or lacking along the coastal plain and in the immediate vicinity of the lower Mississippi River).

FIG. 16.—Cone of Virginia pine × ¾.

Remarks.—This species is one of the four important "southern yellow pines," and much commercial timber is harvested from it. Shortleaf pine like pitch pine is capable of sprouting when young. The taproot is usually curved, and this brings a short section near the surface of the ground. If the tree is injured by fire or otherwise, extra buds on this portion of the root often send up new shoots.

Virginia Pine   Scrub Pine

(*Pinus virginiana* Mill.)

Appearance.—Usually a small tree 30 to 40 ft. in height and 12 to 15 in. in diameter (max. 100 by 3 ft.). The outline is more or less ragged and in general is similar to that of Jack pine. Like several of the other yellow pines, two or three whorls of lateral branches may be produced each growing season.

Needles.—In 2s, 1½ to 3 in. long, rather rigid, grayish green, often twisted and divergent; bundle sheath, persistent.

Flowers.—Male and female occur separately as small cones on the same tree.

Cones require two seasons to mature (for details of growth see description of red pine), 1½ to 2½ in. long, oblong-conical; each scale armed with a sharp prickle and bearing two terminally winged seeds.

Bark.—At first thin and smooth, later broken into reddish-brown, scaly plates.

Habitat.—Found mostly on poor, dry, sandy soils; now spreading rapidly on abandoned farm lands.

Distribution.—From southern New York (Long Island) to southern Indiana, south to northeastern Mississippi and northern Georgia.

Remarks.—In such features as persistence of cones, type of needle, and ability to grow on poor dry soils, this species is similar to the more northern Jack pine. In fact, one may often find short-needled scrub pine that would be difficult to separate from Jack pine. Fortunately from the standpoint of identification, the ranges of these two trees do not meet at any point.

Fig. 17.—Cone of table mountain pine × ¾.

Table Mountain Pine

(*Pinus pungens* Lamb.)

This tree is found in the Appalachian Mountains from south central Pennsylvania southward to northern Georgia. The needles are in 2s, yellow-green, and about 2½ in. long. The most conspicuous feature is the cone whose scales are armed with sharp spikelike tips.

Scotch Pine

(*Pinus sylvestris* L.)

Scotch pine is one of the important timber trees of Europe and many years ago was introduced into our

Fig. 18.—Scot's pine. *1.* Closed cone and foliage × ¾. *2.* Open cone × ¾. *3.* Seed × 1. *4.* Cross section of needle × 35.

country where it is very common in plantations from which it is now escaping by means of seedlings from the old trees. The needles are about 2½ in.

long, in 2s, more or less blue-green, and sharply pointed. The twigs and especially the bark (except on very old trees) is bright orange in color, and the cones have raised scales which make them look something like miniature dinosaurs when closed.

THE LARCHES

Tamarack   American or Eastern Larch

[*Larix laricina* (Du Roi) K. Koch.]

Appearance.—A small to medium-sized tree 40 to 80 ft. high and 1 to 2 ft. in diameter (max. 100 by 2½ ft.), with a long cylindrical trunk and open pyramidal crown.

Leaves are ¾ to 1¼ in. long, linear, soft to the touch, pale blue-green; borne in evident spirals (alternate) on young growth, recurring in tufts on conspicuous dwarfed branchlets (spur shoots) on older growth; deciduous from September to November.

Flowers occur separately as small cones on the same tree; the male yellow, the female bright red.

Cones are ½ to ¾ in. long, erect on the branchlet, with a somewhat oblong outline; scales thin, each bearing two terminally winged seeds.

Twigs.—Brown, marked by many small leaf scars and fewer buds; on older twigs, spur shoots are a conspicuous feature.

Bark.—Thin, grayish to reddish-brown, and minutely scaly.

Habitat.—In the United States, usually restricted to bogs and swamps where it is the most characteristic tree along with black spruce; farther north, it is also found on drier sites.

Distribution.—From Labrador to central Alaska, south to central Minnesota, extreme northern Illinois, Indiana and Ohio, and central Pennsylvania to the

Fig. 19.—Tamarack. *1.* Mature cones $\times \frac{3}{4}$. *2.* Seed $\times 1$.
3. Foliage showing solitary, and clustered leaves $\times \frac{1}{2}$. *4.* Bark
of old tree. *5.* Spur shoot $\times 1\frac{1}{2}$. *6.* First year's twig $\times 1\frac{1}{2}$.

Atlantic Coast (also in extreme western Maryland and adjacent West Virginia).

Remarks.—This is a typical northern tree and together with black and white spruces it reaches the so-called *northern limit of tree growth* in Canada. In that region, growth is extremely slow and "trees" 2 to 3 ft. high may be twenty years or more in age.

The wood is harder and heavier than that of other northeastern conifers, but as firewood it is similar to

other softwoods (see pine, page 34). According to Seton fence posts of this tree last for 20 years, and Emerson mentions its durability in wooden ship construction. Shipbuilders sought especially roots grown at right angles to be wrought into "knees" for joining ribs to deck timbers. Such pieces were found on trees growing on shallow soils underlain by hardpan or flat rock; here the roots, at first growing downward, soon met an obstacle and bent sharply to grow almost horizontally.

Fig. 20.—European larch. *1.* Male flowers × 1. *2.* Female flower × 1. *3.* Open cones × ¾.

Tamarack bark contains tannin which can be extracted and used in tanning leather. The seeds are eaten by several birds including ruffed grouse, and the branchlets sometimes are browsed by deer. Also, the leaves contribute to the diet of the sharp-tailed grouse and snowshoe rabbit. The roots can be used for sewing birch bark and are treated as described under spruce (page 56).

European Larch

(Larix decidua Mill.)

This is a commonly planted European tree differing from tamarack as follows: (1) foliage yellow-green; (2) cones larger, up to $1\frac{1}{4}$ in. long; (3) twigs yellowish and stouter; (4) bark thicker, more plated on older trees.

THE SPRUCES

Leaves.—Persistent for several seasons; needlelike but much shorter than those of pine; borne singly and closely in spirals; four-sided or diamond-shaped in cross section (in eastern species). Each leaf is borne upon a short peglike projection from the twig, and when leaf fall finally occurs, these pegs are left as a characteristic feature of the bare twig, which then looks something like a "sardine's spine" (page 57).

Flowers appear as small cones; male and female on the same tree, the latter often red or purplish and, although small, very striking in appearance. After pollination, they slowly lose their bright color, turn downward, and mature in late summer or early autumn.

Fruit.—A pendent woody cone with thin, smooth or ragged-edged scales.

Remarks.—The spruces produce less commercial timber than the pines but in contrast have furnished most of the pulpwood for newsprint and other purposes. As firewood, spruce is similar to pine (page 34) but not so resinous. The naturally curved branches are useful when placed with the curved side up as "springs" upon which to thatch balsam branchlets to make a bough bed. When of the proper hardness, the resinous exudations from bark wounds are the woodsman's chosen chewing gum. If you try to chew it when too soft, it will stick to every tooth in your head; if too hard, it will break up into a bitter powder when chewed, even though held in the mouth for a while to soften. Like many other accomplishments, a technique is involved!

The long slender roots of the spruces (especially white spruce) are used to sew birch bark in canoe construction. They are gathered, coiled, and placed for about an hour in hot wood ashes to steam. They are then removed, split, and soaked in hot water just before using.

Although widely sold for Christmas trees, spruces have the disadvantage of rapidly losing their leaves when brought into the house.

Spruce seeds of various species are eaten by squirrels and by at least 15 kinds of birds.

Key to the Spruces

1. Cones, 4″ or more in length; needles, diamond-shaped in cross section, difficult to roll between thumb and forefinger....................**Norway spruce** (p. 63)
1. Cones, 2″ or less in length; needles, 4-sided, readily rolled.. **2**

    2. Foliage, yellow-green; cone scales, rounded and smooth or slightly ragged......**Red spruce** (p. 56)
    2. Foliage, blue-green; cone scales, straight and smooth at the tip, or rounded and very ragged..... **3**

3. Twigs, with a grayish bloom, never hairy; bruised foliage, with a pungent sometimes skunklike odor; cone scales, straight and smooth at the end..........
..............................**White spruce** (p. 59)
3. Twigs, more or less finely hairy; foliage, without a rank odor; cone scales, rounded and ragged..............
                     Black spruce (p. 61)

Red Spruce

(*Picea rubens* Sarg., Syn. *P. rubra* Link)

Appearance.—A forest tree 60 to 70 ft. high and 1 to 2 ft. in diameter (max. 120 by $3\frac{1}{2}$ ft.), with a broadly conical crown which in open grown trees reaches nearly to the ground. In the forest, the trunk is tall, straight, and clear of branches and supports a small, broadly pyramidal or rounded crown.

Leaves are ½ to ⅝ in. long, spirally arranged, linear, four-sided, yellow-green, borne on slender

FIG. 21.—Red spruce. *1.* Foliage × ½. *2.* Dead twig showing pegs where leaves grew. × 2. *3.* Closed cone × ¾. *4.* Seed × 1. *5.* Open cone × ¾.

pegs which are a part of the twig.

Flowers occur separately as small cones on the same tree; the male bright red, the female greenish red to pink.

Cones are 1¼ to 2 in. long, pendent, ovoid-oblong; scales, thin and slightly ragged on the margin, each bearing two terminally winged seeds.

Twigs.—Orange-brown, more or less hairy; where the leaves have fallen, conspicuously covered with woody pegs (see "Leaves" above).

Bark.—Thin, grayish to reddish brown, eventually covered with small scales.

Habitat.—Does best on moist, sandy loam soils but also occurs in bogs and on upper, dry rocky slopes.

Distribution.—Nova Scotia, New Brunswick, southern Quebec, and Ontario south through New York and New England to northern Pennsylvania; in the Appalachian Mountains from Maryland to Georgia and west to West Virginia.

Remarks.—This is one of the important forest trees of the northeast and the most common spruce of the Adirondack, Green, and White Mountains. It is cut for lumber and is also the chief source of pulpwood in this region. Spruce wood, because of its resonance, is especially adapted to sounding boards in musical instruments; also food containers made from it impart no unpleasant taste to the contents. With the other spruces, this species is cut for Christmas trees.

FIG. 22.—Old-growth red spruce bark.

Spruce beer, a remedy for scurvy, was made by boiling the young twigs (with their leaves) and adding molasses, honey, or maple sugar. When fermented, it was ready to use.

The tall straight trunks made admirable masts, and crooked roots were wrought into "knees" (see tamarack).

White Spruce

[*Picea glauca* (Moench) Voss.]

Appearance.—An important forest tree, 60 to 70 ft. high and 18 to 24 in. in diameter (max. 120 by 4 ft.). The trunk is long, straight, and tapering and supports a handsome conical crown of blue-green foliage.

Leaves are ⅛ to ¾ in. long, spirally arranged, linear, four-sided, blue-green, with a tendency to mass toward the upper side of the twig; when crushed giving off a strong, sometimes unpleasant odor (hence the name *cat spruce*).

Flowers occur separately as small cones on the same tree, the male yellowish red, the female purplish.

Cones are 1½ to 2 in. long, pendent, narrowly oblong; scales, thin, flexible, smooth and straight on the margin; each bearing two terminally winged seeds.

Twigs—Grayish, often with a pale bloom, otherwise similar to those of red spruce.

Bark.—Thin, ashy brown, scaly, the inner layers often with silvery streaks.

Habitat.—Usually found on moist sandy loam soils, especially along lake shores, stream banks, and adjacent slopes.

Distribution.—Mostly a Canadian species with its northern limit from Labrador to Alaska; in the United States, found in Maine and as far south as northern New Hampshire, Vermont, New York, central Michigan, Wisconsin, and Minnesota.

Remarks.—This is the most important of the eastern and northern spruces in terms of present supply. It is the chief source of pulpwood in northeastern North America and also contributes to the

Fig. 23.—White spruce. *1.* Closed cone and foliage $\times \frac{3}{4}$. *2.* Open cone $\times \frac{3}{4}$. *3.* Bark. (*Photograph by Canadian Forest Service.*)

lumber marketed as spruce (for uses see red spruce). White spruce makes more rapid growth than red or black spruce and is widely planted in reforestation. It also is often used as a Christmas tree. The Indians dug the pliable roots of this tree and used them for lacing their birch bark canoes and for baskets (for details of preparation see spruces, page 56).

Black Spruce

[*Picea mariana* (Mill.) B.S.P.]

Appearance.—A small to medium-sized tree, 30 to 40 ft. high and 6 to 12 in. in diameter (max. 100 by 3 ft.). The trunk is long, straight, and tapering and supports a conical crown of usually short branches.

Leaves are ¼ to ½ in. long, spirally arranged, linear, four-sided, blue-green, persistent on the twig 7

Fig. 24.—Black spruce. *1.* Closed cone × ¾. *2.* Open cone × ¾. *3.* Bark. (*Photograph by Canadian Forest Service.*)

to 10 years.

Flowers occur separately as small cones on the same tree; the male bright red, the female purple.

Cones are ½ to 1½ in. long, pendent, ovoid, purplish the first winter, persistent for many years; scales, thin and ragged on the margin, each bearing two terminally winged seeds.

Twigs—Brownish, hairy, otherwise similar to those of red spruce.

Bark.—Thin, grayish or reddish brown, broken into flaky scales.

Habitat.—In the United States, found mostly with tamarack in bogs; farther north also on better drained, sandy loam soils.

Distribution.—From Labrador to Alaska; in the east, south to New Jersey and west through Pennsyl-

Fig. 25.—Cone and seeds of Norway spruce × ¾.

vania, western New York, southern Michigan, Wisconsin, and Minnesota and northwestward into Canada. (Also in the Appalachians of western Maryland and near the boundary of Virginia and North Carolina.)

Remarks.—Because of its usually small size, this tree is not often cut for lumber. However, it contributes to the supply of pulpwood and is also used

as a Christmas tree. In bogs, the lower branches
often become embedded in the sphagnum moss,
take root, and in time become erect so that they look
like separate trees, a process known as *layering*.
Here growth is so slow that dwarf trees only a few
feet high are found to be twenty to forty years old
and often bear cones profusely—a sign of maturity.
Spruce beer can be made from the young branch
tips (see red spruce).

Norway Spruce
[*Picea abies* (L.) Karst.]

This commonly planted European spruce is similar
in appearance to red spruce except that (1) the leaves
are usually diamond-shaped in cross section (flattened), (2) the branches on older trees are more
drooping, and (3) the cones are 4 to 7 in. long, with
wedge-shaped ragged scales.

Eastern Hemlock[1]
[*Tsuga canadensis* (L.) Carr.]

Appearance.—A forest tree, 60 to 70 ft. high and
2 to 3 ft. in diameter (max. 160 by 5 ft.) with a poorly
pruned trunk (side branches persistent) and an
irregularly conical crown, often somewhat ragged in
outline. Young trees have a dense crown of handsome
foliage, the lower branches reaching to the ground.
Unlike its coniferous associates, hemlock displays a
drooping terminal leader which tends to point away
from the direction of the prevailing wind.

Leaves are ⅓ to ⅔ in. long, spirally arranged but
in flat sprays, linear, dark green, flattened in cross

[1] Carolina hemlock, much used ornamentally, is an
Appalachian tree ranging from Virginia to Georgia. The
sprays are not flattened; the leaves and cones slightly
longer than those of eastern hemlock.

section, rounded at the tip, marked on the under-surface with two white bands, and attached to the twig by a short slender stalk.

Fig. 26.—Eastern hemlock. *1.* Female flower × 2. *2.* Male flowers × 2. *3.* Closed cone × ¾. *4.* Open cone × ¾. *5.* Seed × 1. *6.* Foliage × ½. *7.* Leaf showing white bands below × 2. *8.* Bark of old tree.

Flowers occur separately as small cones on the same tree; the male yellow, the female pink.

Cones are ½ to ¾ in. long, pendent, oblong-ovoid; scales, thin, with smooth margins, each bearing two terminally winged seeds.

Bark.—At first flaky or scaly, soon becoming furrowed, and on old trees divided into rectangular blocks of irregular outline. Freshly cut bark shows purplish streaks.

Habitat.—Found on many types of soil but for best development needs plenty of moisture. Young trees prosper even on old prostrate, partially rotted tree trunks or stumps, and it is not uncommon to find one or more hemlocks and yellow birches competing for the supremacy of a single stump.

Distribution.—Nova Scotia to the north shore of Lake Superior, southward through the Lake states, the northeast, and along the Appalachians to Georgia.

Remarks.—While plenty of white pine was yet available, hemlock was not utilized for lumber because of its brittle, splintery, usually knotty wood. Its chief importance was as a source of tannin (bark). Now, however, considerable lumber is manufactured from hemlock for cheap construction and similar uses. As an ornamental, hemlock has much to recommend it. Young trees are exceedingly graceful; they will grow under considerable shade and may also be trimmed to advantage, the latter an important consideration in using them for hedges, where they are unexcelled. Since the terminal shoot tends to droop toward the east (away from the direction of prevailing winds), it serves with reservations as a "natural compass." The wood of conifers, and hemlock in particular, is noted for its tendency to throw out sparks when used for a campfire. That of hemlock should never be used if a tent is near by or if the fire is made inside a tepee.

Moreover, as Seton says, hemlock knots are "probably the hardest vegetable growth in our woods"; they have a flinty consistency, and to strike one with the blade of a hatchet is to invite a large nick therein. The early settlers made a tea from the twigs and leaves, given to induce sweating. Cloth can be dyed a dull red in an extract of the "pink middle bark" of hemlock. Hemlock tannin was mixed with that from the oaks to avoid a reddish color of the leather, liable to rub off in use.

According to Michaux, the Indians made a poultice for sores and wounds by boiling the inner bark and then pounding it between two stones until it resembled a plaster. White pine bark was used in the same way.

New England housewives made brooms from hemlock branchlets (also a useful thing to have in camp). Hemlock seeds are eaten by several kinds of birds, and deer browse the branchlets in winter. W. L. Webb has observed porcupines cutting branchlets in the tops of forest hemlocks. These fans fall to the ground and appear to attract the deer, which in any event gladly clean them up. This is perhaps one of the few activities of the porcupine that is not a nuisance, and even a detriment to the forest, since he destroys many trees by chewing their bark.

Balsam Fir

[*Abies balsamea* (L.) Mill.]

Appearance.—A small to medium-sized forest tree, 40 to 60 ft. high and 12 to 18 in. in diameter (max. 85 by 2 ft.), with a dense, tapering, spirelike crown which characterizes it from its associates. In the open, the crown reaches to the ground; in the forest, it is restricted and is supported by a long, moderately tapering trunk.

Leaves are ¾ to 1½ in. long, spirally arranged, linear, dark green, flattened in cross section, rounded or slightly notched at the tip, marked below with two white bands; attached directly to the twig without a prominent leaf stem (leaves near the top of the tree

Fig. 27.—Balsam fir. *1.* Foliage × ½. *2.* Mature cone × ½.
3. Dead twig showing smooth leaf scars × 1½. *4.* Cone with
scales partly fallen × ¾. *5.* Cone scales nearly all fallen × ¾.
6. Back of cone scale showing bract × 1¼. *7.* Face of cone scale
showing seeds × 1¼. *8.* Bark of old tree.

usually shorter, sharp-pointed, tending to mass toward the upper side of the twig).

Flowers occur separately as small cones on the same tree; the male yellowish to red, the female with purplish, green-tipped bracts.

Cones are 2 to 4 in. long, upright on the twig, oblong-cylindrical, purplish. Scales fall from the axis at maturity (in this respect differing from all other northeastern trees); each scale bears two seeds with broad terminal wings.

Twigs show plainly small round smooth scars left by the fallen leaves; buds, short and sticky.

Bark.—Smooth, dull green, later with grayish patches, and marked with numerous resin blisters.

Habitat.—Found in moist locations near streams and in swamps; also much dwarfed near mountain tops where grotesque wind forms are often produced.

Distribution.—Principally a Canadian tree from Labrador to the northwestern provinces (Mackenzie and Saskatchewan); in the United States, through Minnesota, Wisconsin, Michigan, west central New York, and central Pennsylvania to the Atlantic Coast (also in mountains of Virginia and West Virginia).

Remarks.—This tree is of little use as a lumber producer because of its soft, weak wood of poor durability, which is, however, mixed with that of the spruces for pulp. Balsam is widely used as a Christmas tree and is superior to the spruces if persistence of leaves is desired after several weeks indoors. The blisters on the bark of standing trees contain a resin that because of its index of refraction, is widely used as a glass cement in optical instruments. It also has medicinal properties in throat afflictions, which were known to the pioneers. According to some writers, the resin was useful in healing cuts and other wounds, but Michaux said it caused inflammation and acute pain; and if given "inconsiderately," it "causes heat in the bladder." Emerson states that a valuable varnish for water colors can be made from the resin. In the woods, the boughs are used for beds, the branchlets for pillows, and the dry wood for fire by friction sets. Josselyn wrote "The Knots of this tree and fat pine are used by the English (colonists) instead of candles, and it will burn a long time, but it makes the people pale."

Balsam seeds are eaten by several birds including the ruffed grouse; and the trees are heavily browsed by deer and moose in winter.

THE "CEDARS"[1]

KEY TO THE CEDARS

1. Fruit, a brown, dry, oblong, woody cone; foliage, yellow-green; branchlets, flattened..................
   Northern white-cedar (p. 69)
1. Fruit, globose, more or less fleshy, bluish or purplish (may be berrylike)............................. **2**

2. Branchlets, 4-sided, fruit a blue "berry"; tree found mostly on dry soils......................
   Eastern redcedar (p. 74)
2. Branchlets, flattened; fruit splitting at maturity, fleshy to leathery; tree, mostly in swamps........
   Atlantic white-cedar (p. 72)

Northern White-cedar (Arborvitae)

(*Thuja occidentalis* L.)

Appearance.—A medium-sized tree, 40 to 50 ft. high and 2 to 3 ft. in diameter (max. 80 by 6 ft.) with a dense pyramidal crown reaching nearly to the ground. In the forest, the crown is more oblong and is supported by a buttressed, sometimes twisted

[1] The true cedars of antiquity belong to another group with leaves like larch, but evergreen.

trunk which may divide into several secondary branches.

Leaves.—Very small and scalelike, giving a jointed appearance to the branchlets, opposite, about ⅛ in. long, dark yellow-green, of two sorts: those on the

Fig. 28.—Northern white-cedar. *1.* Foliage and cones × ¾. *2.* Male flowers before and after shedding pollen × 2. *3.* Female flower × 2. *4.* Open cone × 2. *5.* Seed × 2. *6.* Bark of old tree.

front and back flattened, the ones on the sides boat-shaped (on leading shoots larger and not of two kinds).

Flowers occur separately as very small cones on the same tree; the male yellowish, the female pinkish green.

Cones are ⅓ to ½ in. long, upright on the twig, oblong; scales, rounded or sometimes minutely spine tipped, each scale bearing two or three laterally winged seeds.

Bark.—Thin and shreddy, on old trees forming a fibrous outer network; inner bark, very tough and may be made into string or rope.

Habitat.—Found on many different sites but does best on soils of limestone origin. Common in swamps but also found on dry rocky ridges especially if of limestone.

Distribution.—From Nova Scotia to southeastern Manitoba and southeast through the Lake states including southern New York (but not Pennsylvania) to the coast; also in the mountains of Virginia and North Carolina.

Remarks.—The name *arborvitae*, Latin for *l'arbre de vie* (tree of life), is said to have been given this tree by the king of France in the early sixteenth century. The expedition of Jacques Cartier which wintered on the St. Lawrence River brought this species to France, and it was probably the first American tree to be introduced into Europe. The story (according to Thomas Meehan who published it in 1882) is that members of the crew were cured of "distemper" (scurvy?) by a decoction of evergreen tips (probably of this species) given them by the Indians. Because of this, the tree was brought back as a token of gratitude. In any event, it was soon established in Europe and has been widely distributed by rooted cuttings. They can be produced by anyone who has a box of clean sharp sand and reasonable patience. Cut off several branchlets where they join a larger stem; set them firmly about 2 in. deep in well-packed sand; and keep moist in a shaded place until rooted (4 to 6 weeks or longer). Many of the ornamental varieties may be propagated in this way.

The wood is soft, light, and very durable (cedar fence posts) and is used in boat and canoe construction, for shingles, and for making fire by friction.

As indicated by Gibson, a folded shaving of the wood may be struck repeatedly with a hammer without breaking. This rubbery toughness makes the wood especially useful for canoes, particularly those in river work where rocks cannot always be avoided. Gibson also says that the Indians made canoe slats by pounding the wood with a stone maul until it separated along the annual rings.

The seeds are eaten by a few birds, and the red squirrel, and the branches heavily browsed by deer in winter; rabbits do some damage by chewing the bark of young trees; the foliage is also eaten by rabbits and some game birds.

The Onondaga Indian name for this tree is Oo-sootah, which means feather-leaf. The outer dry bark finely shredded makes good tinder for fire by friction.

Atlantic White-cedar

[Chamaecyparis thyoides (L.) B.S.P.]

As its name indicates, this species is mostly a southern tree. It is found in the northeast only along or near the Atlantic coastal plain from extreme southeastern Maine southward. The foliage is similar in appearance to that of northern white-cedar, but bluish green rather than yellow-green; the cones are quite different, globose, semifleshy, with mushroom-shaped scales. This tree is commonly found in swamps where it often occurs in dense pure stands.

The wood is similar to that of arborvitae. Gibson says that it was the first American wood to be used in pipe organs. Mittelberger is said to have noticed its resonance by listening to the rain pattering overhead upon cedar shingles. He made some organ pipes from it and claimed that this was the best wood that he had found for the purpose.

FIG. 29.—Atlantic white-cedar. *1.* Foliage and cones × ¾. *2.* Closed cone and seeds × 2.

Eastern Redcedar[1]   Redcedar Juniper

(*Juniperus virginiana* L.)

Appearance.—A small to medium-sized tree, 40 to 50 ft. high and 1 to 2 ft. in diameter (max. 100 by 4 ft.), with a dense, narrowly pyramidal or columnar-shaped crown often reaching to the ground.

Fig. 30.—Eastern redcedar. *1*. Juvenile foliage × ¾. *2*. Mature type of foliage and cones × ¾. *3*. Bark of old tree.

Leaves.—Opposite, dark green; on young trees or leading shoots, about ½ in. long, slender, sharp-pointed; on older growth, scalelike and so shaped that the branchlets are four-sided in cross section. Except on very old or slow-growing trees, both types of foliage can readily be found.

[1] Oldfield juniper sometimes reaching tree size is featured by long sharp needles arranged in whorls of three. This species is usually shrubby.

Flowers.—Minute and borne separately as small cones on different trees; male on one tree, female on another.

Fruit is ¼ to ⅓ in. in diameter, a globose, bluish, fleshy, berrylike cone.

Bark.—Thin, reddish brown weathering to gray, fibrous, somewhat flaky on the surface.

Habitat.—Widely spread on many different sites but grows best on soils of limestone origin; commonly found in open stands on poor dry soils throughout its range.

Distribution.—From western Nova Scotia to southeastern North Dakota, south to eastern Texas and Georgia.

Fig. 31.—Open-grown eastern redcedar.

Remarks.—The fruit, available the year around, is eaten by birds (stomach records, 29 species; observations, 52); and in this way, the seeds are widely distributed. Gin presumably gets its name from the fact that in its preparation, alcohol is distilled over juniper "berries" to produce the characteristic flavor.[1] The heartwood is a deep red in color and is used for cedar chests, pencils, and cabinetmaking. Many ornamental varieties are known and widely used in landscape plantings.

[1]"Gin" apparently comes from the French word "genievre" meaning juniper berry.

THE BROADLEAVED TREES OR HARDWOODS

THE WILLOW FAMILY

THE WILLOWS

Leaves.—Simple, alternate, finely serrate, mostly long and lance-shaped; usually with fine hair, a whitish bloom, or both underneath; persistent stipules typical in some species.

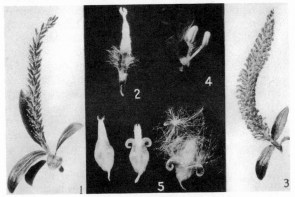

FIG. 32.—Willow. *1*. Female catkin × ¾. *2*. Female flower showing basal nectar gland × 6. *3*. Male catkin × ¾. *4*. Male flower × 6. *5*. Successive stages in the opening of the capsular fruit, with the liberation of a seed × 5.

Flowers.—Male and female borne on separate trees, in more or less erect catkins ("pussies"); individual flowers, very small, equipped with nectar glands.

Fruit.—A small capsule which splits at maturity (usually in late spring) to discharge silky-haired seeds

77

capable of being carried long distances by the wind. The seeds must find a moist place to germinate soon after release, or they die. Placed upon moist paper under a tumbler, they start to sprout within a few hours; and in 2 to 3 days, a pair of tiny seed leaves appears at the top of a delicate stalk.

Twigs.—Usually slender, often highly colored (such as green, red, orange); buds, alternate, each covered with a single caplike scale; terminal bud, lacking.

Remarks.—Willow wood is soft, of little commercial importance except for special purposes. As firewood, (when dry) it produces a quick, bright blaze which soon burns to dead ashes. The bark of some species is tough and can be used as string or twisted into rope (however, basswood and elm are probably better). Willow baskets are made from long shoots, mostly from the introduced purple willow (with subopposite leaves), which has now escaped in some parts of the east and is especially suitable for making willow beds (see Mason's *Woodcraft* for details of construction). In certain localities, honeybees depend upon pollen and nectar from willows for raising spring broods. Willows are readily grown by sticking a live branch into moist soil, whereupon rooting soon takes place. Stream banks are preserved by anchoring mats of willow branches against them; and when rooted, the willows grow and prevent the stream from washing away the soil. Willow bark contains tannin and a bitter principle that has been used as a substitute for quinine. The twigs are eaten by rabbits and beaver, deer and moose; and the buds by 20 or more bird species.

Key to the Willows

1. Upper and lower surfaces, both shiny...............
                                        Shining willow (p. 85)
1. Lower surface, not shiny.......................... **2**

2. Both surfaces, green and smooth; apex, often curved.....................**Black willow** (p. 79)

2. Upper surface, green; the lower, with a gravish or silvery bloom, hairy, or both.................... **3**

3. Buds and leaves, staggered, many appearing paired...
Purple willow (p. 83)

3. Buds and leaves, truly alternate.................... **4**

4. Leaves, lance-shaped......................... **5**

4. Leaves, narrowly elliptical..................... **8**

5. Tree, with a weeping habit; branches pendent.......
Weeping willow (p. 82)

5. Tree, with an upright form....................... **6**

6. Twigs, orange or golden-yellow.................
White or **Golden willow** (pp. 81, 82)

6. Twigs, greenish................................ **7**

7. Leaves, long-stalked; base of twig not brittle........
Peach-leaved willow (p. 82)

7. Leaves, short-stalked; base of twig readily snaps off from larger branch............**Crack willow** (p. 82)

8. Veins, sunken into the upper surface; undersurface, bluish-white-hairy........ .**Bebb willow**   (p. 84)

8. Veins, not sunken; undersurface, with a bluish-white bloom...............**Pussy willow** (p. 84)

Black Willow

(*Salix nigra* Marsh.)

Appearance.—A small to medium-sized tree, 30 to 40 ft. high and 1 to 2 ft. in diameter (max. 100 by 4 ft.), with a short trunk and large irregularly shaped crown. Often more than one trunk arises from the same root crown because of the sprouting habit of this species (also other willows.)

Leaves.—Alternate, simple, 3 to 6 in. long, ⅜ to ¾ in. wide, lance-shaped, smooth and green above and below; the apex, often curved; margin, finely serrate; stipules, on vigorous shoots persistent until autumn.

Flowers.—Male and female borne separately in catkins on different trees.

Fig. 33.—Black willow. *1.* Bark. *2.* Leaf showing semipersistent stipules ("little leaves") × ½. *3.* Fruiting catkin × ¾. *4.* Buds × 1¼.

Fruit.—A small, narrowly ovoid capsule about ¼ in. long, maturing in late spring and opening to set free the minute seeds, each with a parachute of silky hair.

Bark.—Brown to nearly black, heavily furrowed.

Twigs.—Slender, reddish brown to orange-green, brittle; terminal bud, lacking; the laterals, each with a single caplike outer scale; leaf scars, V-shaped, with three bundle scars.

Habitat.—Like the other willows, found usually in moist or wet places such as stream banks, lake shores, and swamps.

Distribution.—From Nova Scotia to northern Minnesota, south to eastern Texas and northern Georgia.

Remarks.—Of all the native willows, this is the only one that can be considered of any commercial importance as a timber producer. The southern variety *altissima* reaches the greatest size and is more frequently cut than the species. Seton says that a decoction of the bark and root is the best substitute for quinine.

OTHER WILLOWS

To use flower and fruit characters of the willows, previous training in botany is almost a necessity, and one must also have a hand lens or low-power binocular microscope. Bark features of the willows are very similar, so that the following descriptions are restricted to leaves and, in some cases, twigs when the coloration is characteristic.

Golden Willow   Golden Osier

[*Salix alba* var. *vitellina* (L.) Stokes]

FIG. 34.— Leaf of golden osier × ½.

Leaves.—Alternate, 2 to 4 in. long, lance-shaped or sometimes broadest near the middle, finely serrate; dark green above and smooth or slightly hairy with a whitish bloom and more or less silky below.

Twigs.—Golden-yellow, later yellowish brown.

Distribution.—Introduced from Europe, now widely naturalized throughout eastern North America.

Remarks.—The white willow *S. alba* L. is less common; the leaves are more silky above than those of golden willow, and the twigs are reddish green.

Weeping Willow

(*Salix babylonica* L.)

Leaves.—Alternate, 3 to 6 in. long, lance-shaped to linear, finely serrate, dark green above, grayish green below.

Distribution.—Originally found in northern China, now widely spread through introduction in Europe and North and South America.

Remarks.—The form of the tree is its most striking feature. With arching limbs and long, pendulous branchlets it presents an unmistakable outline.

Crack Willow

(*Salix fragilis* L.)

Leaves.—Alternate, 3 to 6 in. long, oblong-lance-shaped, finely serrate, dark green above, silvery below.

Twigs.—Brownish, brittle where they join the larger branches.

Distribution.—Found originally in Europe and Asia, now widely naturalized in the eastern United States and Canada; common along watercourses.

Peach-leaved Willow

(*Salix amygdaloides* Anders.)

Leaves.—Alternate, 2½ to 4 in. long, lance-shaped or ovate-lance-shaped, with a fine almost hairlike tip, finely serrate, light green and shiny above, pale below.

Distribution.—Western Quebec to British Columbia, south to central New York, westward to the

Rocky Mountains, and south along their slopes to northwestern Texas.

The following willows although usually shrubby, occasionally reach small tree size.

Purple Willow

(*Salix purpurea* L.)

Leaves.—Alternate or often nearly opposite, 2 to 4 in. long, with almost parallel sides, usually broadest above the middle, finely serrate, dark green above, sometimes with a purplish cast, pale below.

Fig. 35.—
Leaf of purple willow ×
½.

Distribution.—Widely distributed in Europe, Asia, and North Africa; found here as an escape.

Fig. 36.—Making Indian willow bed of purple willow.

Remarks.—This is one of the species introduced for the manufacture of baskets and other willow ware.

Pussy Willow

(*Salix discolor* Muhl.)

Leaves.—Alternate, 1¾ to 4 in. long, more or less oblong but sometimes widest above the middle, finely serrate to nearly entire, dark green above, silvery below.

Distribution.—Nova Scotia to Manitoba, south to Delaware and Missouri.

Fig. 37.— Leaf of pussy willow × ½.

Fig. 38.— Leaf of Bebb's willow × ½.

Bebb Willow

(*Salix bebbiana* Sarg.)

Leaves.—Alternate, 1 to 3 in. long, elliptical or sometimes widest near the tip, finely serrate or entire on the margin, dull green and with prominently sunken veins above, bluish gray and hairy below.

Distribution.—From Newfoundland to Alaska and south to Pennsylvania and Arizona.

Remarks.—This is one of the commonest "weed species" on burned-over forest areas.

According to Seton, it is known also as *withy willow* because of the good withes twisted from it. He says that the inner bark is the best plant material of the north for making fishlines.

Shining Willow

(*Salix lucida* Muhl.)

Leaves.—Alternate, 3 to 5 in. long, narrowly ovate to elliptical or lance-shaped, finely serrate, shiny on both surfaces.

Distribution.—Newfoundland to northwestern Canada, south to New Jersey and Nebraska.

Remarks.—Because of its attractive foliage, it merits consideration as an ornamental.

THE POPLARS

Leaves.—Simple, alternate, usually toothed, triangular to circular; leaf stem in most species flattened, which causes the leaves to quiver in the slightest breeze.

Flowers.—Similar to those of willow, but in drooping catkins and without nectar glands.

Fruit.—A small capsule similar to that of willow, but usually larger, borne in necklacelike strings; seeds like those of willow (page 78) and so small that they also die because of drying out unless they alight upon a moist place soon after release. The capsules contain a large amount of cottony material in addition to that actually attached to the seeds, and this is sometimes a nuisance when poplar trees are used ornamentally.

Twigs.—Moderately slender to stout; buds, alternate; terminal, present; bud scales, several to many; pith, star-shaped.

Remarks.—Poplar wood is soft, not strong and in fire building is similar to that of willow (page 78).

Although it rots quickly when exposed to moisture, George Emerson quotes in connection with poplar wood, "Though heart of oak be e'er so stout, keep me *dry* and I'll see him out." According to the same author, poplar should be good for floors because it dents rather than splinters, can be readily scrubbed white, and, catching fire with difficulty, burns slowly.

Many of the poplars can be readily multiplied by cuttings in a similar manner to the willows (page 78). The buds of some species are important winter food for ruffed grouse. Tannin can be extracted from the bark.

KEY TO THE POPLARS

1. Leaf stem, definitely flattened....................... **2**
1. Leaf stem, more or less rounded in cross section...... **7**

    2. Leaf, whitish-woolly beneath **White poplar** (p. 97)
    2. Leaf, smooth and green......................... **3**

3. Leaves, circular in outline........................ **4**
3. Leaves, triangular or with more or less angled bases; twigs, stout; buds, large and resinous.............. **5**

    4. Teeth, fine; buds, brown and shiny..............
                      Quaking aspen (p. 87)
    4. Teeth, coarse; buds, grayish powdery and dull.....
                      Bigtooth aspen (p. 89)

5. Leaves, triangular; base, flat or slightly heart-shaped; crown of tree, wide-spreading, round in outline.......
                      Eastern cottonwood (p. 91)
5. Leaves, somewhat 4-sided; sides of base slope upward; crown, oblong or very narrow..................... **6**

    6. Leaves, looking as though they had been pulled out sideways; crown, narrow, a tall green column......
                      Lombardy poplar (p. 96)
    6. Leaves, narrower; crown, oblong or narrowly oval..
                      Carolina poplar (p. 95)

7. Buds, sticky-resinous (northern trees)............... **8**
7. Buds, sparingly sticky (southern tree).............
                              Swamp cottonwood (p. 95)

  8. Leaves, smooth, more or less egg-shaped, narrow..
                              Balsam poplar (p. 93)
  8. Leaf stem, usually hairy; leaves, broader and more
    heart-shaped.............**Balm of Gilead** (p. 95)

Trembling Aspen   Quaking Aspen   Popple

(*Populus tremuloides* Michx.)

Appearance.—A small to medium-sized tree 30 to 40 ft. high and 1 to 2 ft. in diameter (max. 100 by 3 ft. in the western part of its range). The trunk is clear of branches and supports a small, rounded, open crown.

Leaves.—Alternate, simple, smooth above and below; blade, 1½ to 3 in. in diameter, nearly circular in outline, serrate with somewhat rounded teeth; leaf stem, about as long as the diameter of the blade, conspicuously flattened.

Flowers.—Male and female borne in separate catkins on different trees.

Fruit.—A small narrowly conical capsule, about ¼ in. long, borne spirally along the catkin axis.

Twigs.—Reddish brown, with star-shaped pith; buds, shiny and dark brown; terminal, present; laterals, incurved.

Bark.—At first and for a number of years, smooth and greenish white; eventually dark brown and furrowed.

Habitat.—Found on a variety of sites but especially typical of burned-over areas where it commonly forms extensive pure stands.

Distribution.—Ranges widely from Labrador to Alaska, south to New Jersey and Nebraska, along the Rocky Mountains to Mexico, and the Sierra

Nevada of California (also in scattered patches on the Appalachian Mountains as far south as Kentucky).

Remarks.—This is one of the most common and characteristic of "north country" trees. It is short lived but forms a temporary cover after forest fires

Fig. 39. ——Quaking aspen.   *1.* Twig $\times 1\frac{1}{4}$.   *2.* Leaf $\times \frac{1}{2}$.
3. Bark of small tree showing the change from the smooth type to older furrowed condition.

and in this way serves to protect other trees which start underneath its moderate shade. After about 20 years, the aspen, which requires full light, becomes crowded and tends to die out; this leaves the more tolerant trees such as birch and maple to form the subsequent forest. The inner bark of aspen, although intensely bitter (extract used as a quinine substitute by the pioneers), is the chosen food of the beaver. The buds are important winter food of the ruffed grouse; the bark and twigs are chewed by snowshoe rabbits; and the twigs or foliage browsed the year round by moose. (These remarks apply as well to bigtooth aspen.)

Sargent says that the wood is the principal fuel of the northern Canadian Indians and that it burns freely even when green; apparently the fact that it was fed continuously into fireplaces of Hudson's Bay Company posts gave him this idea. Seton, however, states that "when green it is so heavy and soggy that it lasts for days as a fire check or back-log." Experience with *any* poplar logs bears out the latter view, although when split and partly dried combustion is much better. When a hot enough fire is raging, *any* green wood will "burn" after the moisture has been cooked out of it!

The Onondaga Indian name for this tree is Nut-Ki-e, meaning "noisy leaf." This is reminiscent of the Greeks who said that poplar leaves were like women's tongues—never still.

Bigtooth Aspen

(Populus grandidentata Michx.)

Appearance.—A small or medium-sized tree, 30 to 50 ft. high (max. 70 ft.) and 1 to 2 ft. in diameter, with a long, clear trunk and small, rounded, irregular crown.

Leaves.—Alternate, simple, nearly circular in outline; smooth above and below; the blade, $1\frac{1}{2}$ to $2\frac{1}{2}$ in. in diameter, serrate with coarse teeth; leaf stem, about as long as the diameter of the blade, conspicuously flattened.

Flowers.—Male and female borne in separate catkins on different trees.

Fruit.—A small, narrowly conical capsule about $\frac{1}{4}$ in. long, borne spirally on the catkin axis.

Twigs.—Brownish gray, with star-shaped pith; buds, dull, grayish; terminal, present; laterals, divergent.

Bark.—At first smooth and olive-green, later brownish and furrowed.

Habitat.—Found on widely different sites, but most common following fires, and mixed with quaking aspen.

Fig. 40.—Bigtooth aspen. *1.* Bark of old tree. *2.* Leaf × ½.
3. Twig × 1¼.

Distribution.—Nova Scotia to northeastern North Dakota, through the Lake states, the northeast, and along the Appalachians to Tennessee.

Remarks.—This tree is less common than trembling aspen and in the northeast often reaches greater size. It plays the same role as that species in the growth of new forests following fire. The peeled logs of this and trembling aspen (creosote surface-stained) have been used successfully in log cabin construction (see poplar for durability).

Eastern Cottonwood

(*Populus deltoides* Marsh.)

Appearance.—A medium-sized to large tree, at its best 80 to 100 ft. high and 3 to 4 ft. in diameter (max. 175 by 11 ft.). The crown, although somewhat pyramidal at first, soon becomes rounded and on old trees may be very wide-spreading (80 to 100 ft. across).

Leaves.—Alternate, simple, triangular in outline; smooth above and below; blade, 2 to 4 in. long; margin, with rounded teeth; leaf stem, about as long as the blade, conspicuously flattened.

Flowers.—Male and female borne in separate catkins on different trees.

Fruit.—A small, ovoid capsule, about ⅓ in. long, borne alternately along the axis (looking something like a string of beads, hence, perhaps, the name *necklace poplar* sometimes applied to this tree).

Twigs—Yellowish brown, stout; pith, 5-angled; buds, brownish, sticky; terminal, present, ovoid or widest near the middle.

Bark.—Greenish yellow on young stems, later gray and deeply furrowed.

Habitat.—A common stream-bank tree throughout its range, although prospering on drier sites when once planted.

Distribution.—(Including the southern and western varieties) southern Quebec to Montana, south to west central Texas; and through western Massachusetts southward to northern Florida (rare or absent in most of New England and also the Appalachians).

Remarks.—Like many of the willows and other poplars, this tree is easily propagated by planting a

For descriptive legend see opposite page.

piece of branch in moist earth. If done in the spring, the new sprout may attain a height of 4 to 6 ft. the first season. The Plains Indians used the root wood for starting fire by friction, but as firewood (when dry) it burns quickly without leaving a bed of coals. According to a legend, the Indian discovered the design for his tepee by twisting a cottonwood leaf between his fingers, in this way producing the conical pattern. Indian children still fashion play tepees in this way.

This tree is a year-round larder for the sharp-tailed grouse which has been reported to eat the buds, leaves, and catkins.

Balsam Poplar

(*Populus tacamahaca* Mill.;

P. balsamifera Dur.)

Appearance.—A medium-sized tree, 60 to 80 ft. high and 1 to 2 ft. in diameter (max. 100 by 5 ft.); with a narrow, open, pyramidal crown and long, cylindrical trunk.

Leaves.—Alternate, simple, 3 to 6 in. long, ovate, smooth above, smooth and often with rusty blotches below; margin, finely serrate, with rounded teeth; leaf stem, circular in cross section.

Flowers.—Male and female borne in separate catkins on different trees.

Fruit.—A small, ovoid to oblong capsule about $\frac{1}{4}$ in. long borne alternately along the axis.

Twigs.—Reddish brown, shiny, with star-shaped pith; buds, reddish brown, sticky; terminal, present.

Fig. 41.—Eastern cottonwood. *1.* Twig $\times$ $1\frac{1}{4}$. *2.* Leaf $\times$ $\frac{1}{2}$. *3.* Male catkins $\times$ $\frac{1}{2}$. *4.* Male flowers, front and rear views, respectively $\times$ 3. *5.* Female flower and bract $\times$ 3. *6.* Fruit clusters $\times$ $\frac{2}{3}$. *7.* Bark, showing lack of butt swell because of frequent inundation and deposition of river silts.

Bark.—Greenish to reddish brown, later becoming grayish, heavily furrowed and with somewhat scaly ridges.

Habitat.—Most common on alluvial bottom lands and along streams or lake shores.

Fig. 42.—Balsam poplar. *1*. Bark. *2*. Leaf × ½. *3*. Fruit × 2. *4*. Twig × 1¼.

Distribution.—Balsam poplar is mostly a Canadian species with its northern limit from Labrador to the coast of Alaska; in the United States, from central Maine to Michigan, westward through South Dakota to the Rocky Mountains of Colorado and through northern Nevada and Oregon to the Pacific Coast.

Remarks.—This is a typical Canadian tree and reaches its best development in the far northwest in the valley of the Mackenzie River. On the northern shores of the Great Lakes, the bark near the base of old trees is sometimes used for fish-net floats.

Balm-of-Gilead

(*Populus candicans* Ait.)

This species is similar to balsam poplar and according to some writers is a variety of it. However, the leaves are more heart-shaped and have coarser teeth and hairy leaf stems. Although most authorities agree that only female flowers are produced (thus suggesting a hybrid origin), at least one writer states that both male and female occur. Where the tree originated is not known, but it is commonly planted in Canada and the northern United States. The resin from the buds contains certain substances used in medicine.

Swamp Cottonwood

(*Populus heterophylla* L.)

Swamp cottonwood is essentially a tree of the southeast and Mississippi Valley but comes north along the coast as far as southern Connecticut (also said to occur in southern Michigan). The leaves are ovate, 4 to 8 in. long and have circular leaf stems (cross section). As indicated by its name, this tree is found on low, swampy ground which may be inundated for considerable time each year.

Carolina Poplar

[*Populus canadensis* var. *Eugenei* (Simon-Louis) Schelle]

The name of this tree is entirely misleading, since it apparently originated in the nursery of Simon Louis near Metz, France, in 1832. The original tree was measured in 1913 (81 years later) by Henry who found it to be 150 ft. high and nearly 8 ft. in diameter. Presumably a hybrid, with the Lombardy poplar as one parent, this tree produces only male flowers. Since there are no seeds, all subsequent individuals after the first have been grown from rooted cuttings. The leaves are 2 to 6 in. long, rhombic-ovate in shape, with flattened leaf stems. They bear some resemblance to those of both eastern cotton-

Fig. 43.—Leaves of white poplar $\times \frac{1}{2}$.

wood and Lombardy poplar but have a more sloping base than the former and are not so wide in proportion as the latter.

At one time because of its rapid growth, Carolina poplar was widely planted in this country as a street tree; but experience has shown that the roots often grow into near-by sewers and plug them, thus causing expensive repairs. In most cities, these trees are being cut down on this account. They should be girdled in the spring and not removed until fall; if this precaution is not taken, numerous sprouts may come up from various parts of the root system.

White Poplar

(*Populus alba* L.)

This is a tree introduced from Europe, with leaves similar in shape to those of our native aspens. The undersurface is more or less silvery-woolly and very striking in appearance as contrasted with the dark green of the upper surface. The variety *nivea* displays leaves of various degrees of lobing, the more extreme ones maplelike in outline. Like the Carolina poplar, this tree is difficult to kill, and often one will see abandoned farmhouses with thickets of young root sprouts coming up from the lawn—all originating from a single specimen planted many years before as an ornamental.

THE WALNUT FAMILY

THE WALNUTS

Leaves. —Compound, alternate; leaflets, lance-shaped to oblong-elliptical, serrate, more or less fragrant when crushed.

Fig. 44.—Walnut. *1.* Female (upper), and male (lower) flowers × ½. *2.* Male flower (face view) × 3. *3.* Female flower × 3. *4.* Male flower (back view) × 3.

Flowers.—Male and female borne on the same tree; the former, in pendent catkins; the latter, in small inconspicuous clusters.

Fruit.—A large bony nut, with an outer fleshy layer that shrivels rather than splits when dry; seed, more or less oily, sweet and edible.

Twigs.—Stout, buds, alternate; terminal, present; leaf scars, large and conspicuous; pith, when cut lengthwise, shows cross partitions separating small cavities (chambered). Between each season's growth is a "plug" of solid tissue; and by counting these, the age of a twig can be determined; in some species, the pith tissue contains beautiful rosette-shaped crystals which look like stars when viewed with a microscope using polarized light.

Remarks.—The fruit of the walnuts including butternut is too well known to discuss at length; it is, of course, an important winter food for squirrels. The bark of certain species contains dye substances used by the pioneers in dyeing homespun cloth. Since the wood of the two native (eastern) species is quite different, it is not mentioned here.

KEY TO THE WALNUTS

1. Leaflets, somewhat parallel-sided; pith, chocolate colored; nut, ellipsoidal . . . .**Butternut** (p. 101)
1. Leaflets, broadest near base; pith, light yellowish gray; nut, spherical . . . .**Black walnut** (p. 98)

Black Walnut

(Juglans nigra L.)

Appearance.—A medium-sized to large tree, 70 to 90 ft. high and 2 to 3 ft. in diameter (max. 150

Fig. 45.—Black walnut. *1.* Leaf × ⅙. *2.* Bark. *3.* Fruit × ¾. *4.* Nut × ¾. *5.* Cross section of fruit × ¾. *6.* Twig × 1¼.

For descriptive legend see opposite page.

by 6 ft.), with a tall trunk, which in the forest is clear of branches; and a small open crown.

Leaves.—Compound, alternate, 12 to 24 in. long, with 15 to 23 ovate-lance-shaped leaflets (often with an even number of leaflets because the terminal one fails to develop); leaflets, finely serrate, smooth above, hairy below.

Flowers.—Borne separately on the same tree; the male in catkins, the female in short spikes.

Fruit.—A large globose "nut" (technically of the nature of a drupe), 1½ to 2 in. in diameter; the outer layer more or less fleshy, the inner, hard and bony and marked with rounded ridges; seed, sweet and edible.

Twigs.—Brownish, stout, with buff-colored, chambered pith; terminal bud, present, short and blunt, larger than the laterals; leaf scars, large and conspicuous, each, with three U-shaped bundle scars.

Bark.—Brown to nearly black, soon divided into narrow furrows; on older trees, deeply furrowed; the fissures separated by sharp ridges; the whole presenting a somewhat diamond-shaped pattern.

Habitat.—Does best on deep, rich, moist soils of alluvial origin; on poorer sites makes relatively slower growth.

Distribution.—From western Massachusetts to central Nebraska, south to eastern Texas and southern Georgia.

Remarks.—Black walnut is one of our finest hardwoods and is widely used for furniture and cabinetwork of all kinds. So scarce has it become that during the First World War, surveys were made to find even individual trees for making gunstocks and airplane propellers. The roots of black walnut are said to give off a substance poisonous to certain plants such as tomatoes. When used as a dooryard tree, it should be remembered that the fruit husks contain a dye capable of staining cloth (so used by the pioneers). Probably few people will have occasion to build the campfire of black walnut, but it will produce a bed of hot coals like other heavy hardwoods.

Butternut   White Walnut

(*Juglans cinerea* L.)

Appearance.—A small to medium-sized tree, 40 to 60 ft. high and 1 to 2 ft. in diameter (max. height 100 ft.), with a short trunk and open scraggly crown, often featured by several dead branches.

Leaves.—Compound, alternate, 15 to 30 in. long, with 11 to 17 oblong-lance-shaped leaflets which are finely serrate on the margin; veins, somewhat sunken on the upper surface; lower surface, hairy. Unlike black walnut, the leaves always have a terminal leaflet.

Flowers.—Borne separately on the same tree; the male in catkins, the female in short spikes.

Fruit.—A large ovoid "nut" (technically a drupe), $1\frac{1}{2}$ to $2\frac{1}{2}$ in. long; the outer layer, somewhat fleshy and covered with sticky hairs, the inner, hard and bony and marked with sharp ridges; seed, sweet and edible.

Twigs.—Greenish brown, stout, with chocolate-colored, chambered pith; terminal bud, present, longer than in black walnut, larger than the laterals; leaf scars, large and conspicuous, each with three U-shaped bundle scars.

Bark.—Ashy gray, at first smooth, soon divided into broad ridges, eventually deeply furrowed, the furrows separated by narrow flat-topped ridges.

Habitat.—Does best on deep, moist, rich soils, but also commonly found on limestone outcrops.

Distribution.—Southern New Brunswick to southeastern South Dakota, south to eastern Arkansas

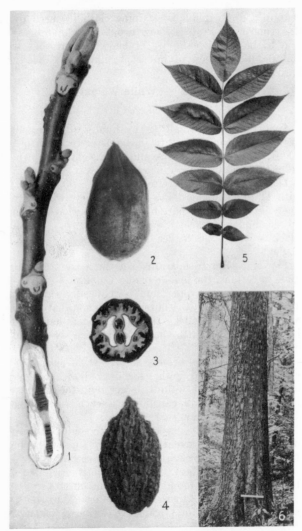

For descriptive legend see opposite page.

and northern Georgia (not on coastal plain below Delaware).

Remarks.—Butternut is of secondary value as a timber tree, since the form is poorer and the wood softer and weaker than that of black walnut. The husks and bark contain a water-soluble yellow dye used by the pioneers for dyeing cloth. As seen in cross section, the nut (also other walnuts and hickories) presents a design that might well be used in the arts and crafts as the basis for such woodcraft items as seals and watch fobs. Cross sections of the nuts themselves make attractive buttons.

Michaux says that the early settlers plunged the half-grown fruits into boiling water, wiped off the sticky down, and pickled them in vinegar. Roger Williams reported of the Indians that "of these Wallnuts they make an excellent oyle good for many uses, but especially for their annoynting of their heads" (probably black walnut as well as some of the hickories are also meant). The pioneers made a mild cathartic by boiling the root bark (1 lb. in a gallon of water boiled down to a quart) and adding honey.

THE HICKORIES

Leaves.—Compound, alternate, usually with fewer leaflets than the walnuts; leaflets, lance-shaped to elliptical or oblong, serrate, more or less fragrant when crushed.

Flowers.—Male and female borne on the same tree; the former in conspicuously three-branched catkins, the latter in small inconspicuous clusters.

Fruit.—A bony nut, usually whitish on the surface, enclosed in a husk that splits and usually falls apart

---

Fig. 46.—Butternut. *1.* Twig × 1¼. *2.* Fruit × ¾. *3.* Cross section of fruit. *4.* Nut showing sharp corrugations × ¾. *5.* Leaf × ⅕. *6.* Bark.

when ripe; seed, sweet and edible or bitter, depending on the species.

Twigs.—Moderately stout to slender, tough and easily tied into knots without breaking; buds, alternate; terminal, present; leaf scars, conspicuous; pith, five-angled in cross section, solid in contrast to that of walnut (page 97).

Remarks.—Hickories are essentially American trees. Of the 18 or 20 species, one is found in Mexico, one in China, one in Indochina, and the rest in eastern United States.

Hickory wood[1] is noted not only for its hardness but also for its toughness and ability to stand up under sudden shocks. For this reason, it has held first place in the making of axe handles and formerly wheel spokes for automobiles. American axes are said to be known around the world not so much for their steel as for their hickory handles. It has also been claimed that the sport of trotting horses was developed in this country because of the invention of the light sulky, in the making of which hickory is indispensable. Although the imported lemonwood is widely used for

Fig. 47.—Hickory. *1.* Flowers showing 3-branched habit of aments × ½. *2.* Female flowers × 3. *3.* Male flower × 3.

[1] These remarks apply particularly to the wood of the "true hickories." That of the pecan group including bitternut is inferior by comparison.

bows, a good stick of hickory cannot be considered so greatly inferior. Hickory also ranks first in fuel value of the common American woods; used in the campfire, it leaves a hot bed of coals ideal for broiling. Good charcoal can be made from it, and the ashes are useful for soapmaking.

From the crushed nuts the Indians made an oil that an early writer said was beneficial for "Dolors

Fig. 48.—Sighting along a hickory axe handle to see if handle and blade are in line. This is necessary for best performance.

and gripes of the Belly." Another wrote that venison broth was thickened and made more savory by adding the pulverized nuts (shells and all) of certain hickories; the shells sank to the bottom, and the soup could be poured off. "Powcohiccora," a fermented drink, was made by pounding the nuts and letting the mash "set" for a while.

Hickory nuts have been found in the stomachs of several of the larger birds (grouse, turkey, ring-necked pheasant, wood duck).

1. Buds, bright sulfur-yellow, long and narrow; husk of nut, winged.....................**Bitternut** (p. 114)
1. Buds, green or brown, ovoid to globose; husk, not winged... **2**

   2. Husk of nut, thick (see p. 107)................. **3**
   2. Husk, thin (see p. 110)........................ **5**

3. Outer bud scales soon fall off leaving a smooth-looking bud; leaves, fragrant, hairy or woolly; nut, 4-sided...
                                      Mockernut (p. 112)
3. Outer scales, persistent; bud, rough and scaly; leaves, slightly fragrant, smooth or velvety; nut, flattened... **4**

   4. Leaflets, 5, mostly smooth; entire fruit, commonly about $1-1\frac{1}{2}''$ in diameter.......**Shagbark** (p. 106)
   4. Leaflets, mostly 7, velvety below; fruit, often up to $2''$ or more in diameter; twigs, with orange spots...
                                Shellbark (p. 109)

5. Husk, clings to nut; leaflets, mostly 5...**Pignut** (p. 112)
5. Husk, falls away; leaflets, often 7 (also 5)...........
                                 Red (p. 111)

Shagbark Hickory

[*Carya ovata* (Mill.) K. Koch.]

Appearance.—A medium-sized forest tree, 70 to 80 ft. high and 1 to 2 ft. in diameter (max. 120 by 4 ft.), with a straight, cylindrical trunk and more or less oblong crown, a feature characteristic of several of the hickories.

[1] Species not included are either essentially southern in distribution or are rare and local.

FIG. 49.—Shagbark hickory. *1.* Twig $\times 1\frac{1}{4}$. *2.* Leaf $\times \frac{1}{4}$. *3.* Fruit $\times \frac{3}{4}$. *4.* Nut $\times \frac{3}{4}$. *5.* Cross section of nut $\times \frac{3}{4}$. *6.* Bark.

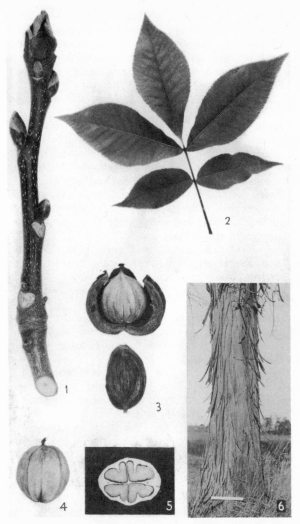

For descriptive legend see opposite page.

Leaves.—Compound, alternate, 8 to 14 in. long, with five (rarely seven) somewhat elliptical leaflets often broadest, however, near the tip; smooth above, smooth or finely hairy below; the upper three leaflets larger than the lower two; margins, finely serrate.

Flowers.—Borne separately on the same tree; the male in three-branched catkins, the female in short spikes.

Fruit.—A bony nut encased in a husk from $\frac{1}{4}$ to $\frac{1}{2}$ in. thick, the whole measuring from 1 to $2\frac{1}{2}$ in. in diameter. The husk splits cleanly to release the nut which is flattened and globose and bears four ribs, two on each side; seed, sweet and edible.

Twigs.—Reddish brown, stout, with star-shaped or five-angled pith; terminal bud, present, larger than the laterals; leaf scars, large and conspicuous; bundle scars, numerous.

Bark.—At first smooth and gray; later, breaking up into characteristic plates the ends of which curve away from the trunk.

Habitat.—Does best on deep, moist, but well-drained alluvial soils.

Distribution.—From southern Maine to eastern Nebraska, south to northeastern Texas and northwestern Florida (below Delaware not on the Atlantic or Gulf coastal plain).

Remarks.—This is one of the best commercial hickories. It produces a heavy, hard, strong wood well known for its ability to withstand sudden shocks, hence formerly used for automobile wheel spokes, and almost universally for such items as axe helves. As mentioned previously, hickory firewood stands first among common North American woods. It not only has the highest heat value but produces excellent coals for baking or broiling. There are several varieties of shagbark hickory, the most widely distributed being littlenut shagbark (*Carya ovata* var.

nuttallii Sarg.) with a small fruit usually less than 1 in. in diameter.

Shellbark Hickory

[*Carya laciniosa* (Michx. f.) Loud.]

This tree is one of the central states' hardwoods, most common in the Ohio River Valley on wet soils

Fig. 50 —Shellbark hickory. *1.* Twig × 1¼.
2. Side view of husk × ¾. *3.* Nut × ¾.

and reaching into the northeast only to central New York. The leaves are 15 to 22 in. long and usually have seven (occasionally five or nine) leaflets, brownish velvety on the lower surface. The fruit looks like an enormous shagbark hickory nut (up to 2½ in. in diameter including the husk), although that species at its best produces nuts almost as large. The terminal buds of shellbark hickory are larger than

For descriptive legend see opposite page.

those of the preceding species, and the twigs bear numerous orange spots (lenticels). The bark plates tend to be flat at their free ends, instead of curving.

Found from central New York to southeastern Nebraska, south to eastern Oklahoma, and along the Appalachians to northern Louisiana.

Red Hickory   Oval Pignut Hickory

[*Carya ovalis* (Wang.) Sarg.]

Appearance.—A medium-sized forest tree, 50 to 70 ft. high and 1 to 2 ft. in diameter (max. 100 by 3 ft.), with a tall trunk and somewhat open, oblong crown.

Leaves.—Compound, alternate, 6 to 12 in. long, with seven (more rarely five), narrowly elliptical to elliptical-lance-shaped leaflets; serrate on the margin; smooth above and below.

Flowers.—Borne separately on the same tree; the male in three-branched catkins, the female in short spikes.

Fruit.—An oval bony nut encased in a husk about $\frac{1}{10}$ in. in diameter; the whole from 1 to $1\frac{1}{4}$ in. in diameter. Husk splits cleanly all the way to the base. Seed is small, sweet or slightly bitter.

Twigs.—Reddish brown, moderately stout, with star-shaped or five-angled pith; terminal bud, present and nearly globose, larger than the laterals; leaf scars, large and conspicuous, bundle scars, numerous.

Bark.—Grayish, at first smooth, soon breaking up into plates not unlike those of shagbark but usually not so wide; on old trees, with rough interlacing ridges, scaly on the surface.

Habitat.—Found mostly on dry upland soils with other hardwoods, especially oaks.

Fig. 51.—Red hickory. *1.* Twig × 1¼. *2.* Leaf × ⅙. *3.* Fruit × ¾. *4.* Cross section of nut × ¾. *5.* Nut × ¾. *6.* Bark.

Distribution.—From central New York to Iowa, south to Arkansas and northern Georgia (not on the coastal plain below the District of Columbia).

Remarks.—This tree is often mistaken for pignut hickory from which it differs chiefly in its fruit and more shaggy bark. A number of varieties have been described principally on the basis of fruit size and shape, and one or more of these reach as far east as New England. This hickory is one of the important commercial species, and the wood has similar properties to those already mentioned under hickories.

Pignut Hickory

[*Carya glabra* (Mill.) Sweet.]

This species is quite similar to oval pignut hickory except that (1) the leaves usually have five (less

FIG. 52.—Fruit of pignut × ¾; note incomplete splitting.

commonly three or seven) leaflets, (2) the bark is less shaggy with rounder ridges, and (3) the fruit is pearshaped and the husk splits only about halfway down. This last feature is the only trustworthy one, since the other characters intergrade. Pignut is one of the commercial hickories (see Shagbark for properties of the wood) and ranges from southern Vermont to southern Michigan, south to Louisiana and Florida. A large-fruited form, var. *megacarpa*, is most common in the south.

Mockernut Hickory

(*Carya tomentosa* Nutt.)

This hickory, although rare in the north, is a common "central hardwood" of the Ohio Valley and

FIG. 53.—Mockernut hickory. *1.* Leaf $\times \frac{1}{4}$. *2.* Twig $\times 1\frac{1}{4}$. *3.* Nut $\times \frac{3}{4}$. *4.* Cross section of nut $\times \frac{3}{4}$. *5.* Fruit $\times \frac{3}{4}$. *6.* Bark.

southward, found mostly on dry upland soils. It may be distinguished as follows: (1) The leaves have seven to nine fragrant leaflets densely hairy underneath and along the main leaf stem; (2) the nut is four-angled and less flattened than those of the shagbarks previously described, with a husk about $\frac{1}{4}$ in. thick, splitting nearly to the base; (3) the terminal bud is similar to that of shagbark hickory but more globose, with outer scales that soon fall off, leaving it smooth during the winter; (4) the bark displays a wavy pattern of shallow furrows and low rounded ridges. Mockernut is one of the commercial hickories (see hickory for uses of wood) and ranges from Massachusetts to northeastern Kansas, south to eastern Texas and central Florida. The Indians made a black dye from the bark of some hickory (probably this one) in combination with a vegetable acid.

Bitternut Hickory

[Carya cordiformis (Wang.) K. Koch.;
Hicoria cordiformis (Wang.) Britt.]*

Appearance.—A medium-sized tree, 50 to 60 ft. high and 1 to 2 ft. in diameter (max. 100 by 3 ft.), with a tall, clear trunk and rounded crown.

Leaves.—Compound, alternate, 6 to 10 in. long, with 7 or 9 (rarely 11) narrowly elliptical to lance-shaped leaflets, serrate on the margin, smooth above, more or less hairy below.

Flowers.—Borne separately on the same tree; the male in three-branched catkins, the female in short spikes.

Fruit.—A nearly globose thin-shelled nut encased in a thin husk which, along the lines of separation, projects in the form of a ridge or wing from the apex halfway to the base of the fruit; seed, bitter.

Fig. 54.—Bitternut hickory. *1.* Twig × 1¼. *2.* Leaf × ⅓. *3.* Fruit × ¾. *4.* Nut × ¾. *5.* Cross section of nut showing thin shell × ¾. *6.* Bark.

Twigs.—Grayish brown, moderately stout, with star-shaped or five-angled pith; terminal bud, present, larger than the laterals, bright yellow, with scales that meet in a vertical line without overlapping; leaf scars, large and conspicuous; bundle scars, numerous.

Bark.—Grayish, smooth for many years (more so than the other hickories of this region) eventually with narrow fissures separating low, firm, interlacing ridges.

Habitat.—Widely distributed, but most typical of moist or wet bottom lands and along streams.

Distribution.—From southern Maine to Minnesota, south to eastern Texas and northern Florida (not on the coastal plain below Virginia).

Remarks.—This is the only representative of the pecan-hickory group found in the northeast; the wood, although utilized, is inferior to that of the "true hickories."

THE BIRCH FAMILY

THE BIRCHES

Leaves.—Simple, alternate, usually doubly toothed, elliptical, or oval to triangular.

Flowers.—Male and female borne in separate catkins on the same tree. The former can be seen all winter (preformed); the latter emerge in spring.

Fruit.—A minute two-winged nutlet ("seed") borne in great numbers in a cone, the scales of which fall apart at maturity. Birch seeds will not stand much drying out, and young trees are therefore typical of moist places. The seeds when first shed require temperatures as high as 90°F. to induce sprouting; but after lying at winter temperatures for several months, they will in spring sometimes germinate in melting snow, a feature not common in woody plants.

Twigs.—Slender, usually zigzag; buds, alternate; terminal bud, lacking; bud scales, several; dwarf, slow-growing branchlets called *spur shoots* common.

Remarks.—Several of the birches produce wood of commercial value. As firewood, it varies with the species; that of yellow and black birch produces better live coals than that of white and gray. Birch seeds are eaten by birds; the buds are an important winter food of grouse and other large birds. Deer, moose, and rabbits browse the twigs in winter, and beaver use the inner bark for food when poplars (aspen) are not available.

KEY TO THE BIRCHES

1. Twigs, with a wintergreen flavor where chewed; leaves, more or less elliptical...................... **2**
1. Twigs, not wintergreen flavored; leaves, oval, triangular, or diamond-shaped........................... **3**

    2. Leaves, doubly serrate, more or less rounded at the base; bark, bronze or yellowish on young trees....
               Yellow birch (p. 118)
    2. Leaves, singly serrate, heart-shaped at the base; bark, black, like that of black cherry............
               Sweet birch (p. 120)

3. Leaves, diamond-shaped; bark, on young trees pinkish, papery; fruits in the spring.. **River birch** (p. 127)
3. Leaves, oval or triangular; bark on trees, over 2″ in diameter, whitish, papery; fruits in the fall.......... **4**

    4. Leaves, oval............. **Paper birch** (p. 123)
    4. Leaves, triangular or rounded-triangular.......... **5**

5. Bark, creamy-white; leaves, rounded-triangular......
               European white birch (p. 128)
5. Bark, dirty or grayish white, black triangular patches beneath the branch insertions common; leaves, triangular with a long narrow tip....**Gray birch** (p. 126)

Yellow Birch

(*Betula lutea* Michx. f.)

Appearance.—A medium-sized forest tree, 60 to 70 ft. high and 1 to 2 ft. in diameter (max. 100 by 4 ft.). In the forest, the trunk is relatively clear of branches and supports an irregular crown; the roots are shallow and wide-spreading.

Leaves.—Alternate (on spur shoots appearing opposite or whorled), simple, 3 to 4 in. long, 1 to 2 in. wide, ovate to oblong-elliptical, sharply doubly serrate, smooth above, sometimes tufted hairy below, in the vein axils.

Flowers.—Male and female borne in separate catkins on the same tree; male catkins present during the winter, female developing from buds the following spring.

Fruit.—A minute winged nutlet (seed) borne in large numbers in an ovoid cone whose scales fall tardily from the axis at maturity; cone upright, about $1\frac{1}{2}$ in. long.

Twigs.—Greenish brown, slender, aromatic (oil of wintergreen); terminal bud, lacking except on spur shoots which are conspicuous features.

Bark.—At first, bronze in color, thin, readily peeling in papery curls; later, changing to coarse scaly plates.

Habitat.—A cool, moist site is most typical for this tree.

Distribution.—The Lake states, southern Canada, the northeast, and in the Appalachians to northern Georgia.

Remarks.—This tree is the most important of the commercial birches and probably furnishes three-fourths of the lumber marketed under the name *birch*. The bark curls are inflammable and even on a wet day in the woods make ideal tinder for starting a fire.

Fig. 55.—Yellow birch. *1.* Twig × 1¼. *2.* Spur shoot × 1¼. *3.* Fruiting cone × ¾. *4.* Fruit × 3. *5.* Bract, or cone scale × 3. *6.* Leaf × ½. *7.* Bark of young tree. *8.* Bark of old tree.

Discretion should be used, however, not to mar the appearance of trees along trails or in public parks. The seedlings of this tree occur in great numbers, wherever sufficient moisture is present. This accounts for the young trees that are often found on the tops of old stumps. In time, the latter often rot away leaving the trees on "stilts" (roots that have meantime grown along the stump until they have reached the ground). In such places, the most common coniferous associate is hemlock whose seedlings will also grow almost anywhere in the forest provided there is abundant moisture.

The seeds of yellow birch and also sweet birch are probably more important as late winter food for birds than those of the white and gray birches, because the cones of the latter fall to pieces within a few weeks after they ripen in the fall whereas those of yellow and sweet birch disintegrate slowly throughout the winter, meantime releasing their seeds which come to rest on top of the snow blanket where they are available. In late March after a windstorm, numerous seeds have been seen cast in this way, and the presence of bird tracks indicated the use of these seeds for food.

Sweet Birch   Black Birch

(*Betula lenta* L.)

Appearance.—A medium-sized forest tree, 50 to 60 ft. high and 1 to 2 ft. in diameter (max. 80 by 5 ft.), with a somewhat tapering trunk and narrow, rounded crown.

Leaves.—Alternate (on spur shoots appearing whorled or opposite), simple, $2\frac{1}{2}$ to 5 in. long, $1\frac{1}{2}$ to 3 in. wide, ovate to oblong-ovate, sharply singly serrate or inconspicuously doubly serrate, more heart-shaped at the base than those of yellow birch, smooth above, often with tufts of hair in the vein axils below.

Flowers.—Male and female borne in separate catkins on the same tree; preformed male catkins present during the winter.

Fruit.—A minute winged nutlet (seed) borne in large numbers in an oblong-ovoid cone whose scales

Fig. 56. ——Sweet birch. *1.* Bark (*Photograph by H. P. Brown*). *2.* Bud × 1¼. *3.* Spur shoot × 1¼. *4.* Leaf × ½. *5.* Fruiting cone × ¾. *6.* Fruit × 3. *7.* Bract, or cone scale × 3.

fall tardily from the axis at maturity; cone, upright, about 1½ in. long.

Twigs.—Dark brown, slender, aromatic (oil of wintergreen); terminal bud, lacking except on spur shoots which are conspicuous features; lateral buds, sharp pointed (more so than those of yellow birch).

Bark.—Nearly black (resembling that of black cherry), at first smooth, eventually scaly plated.

Habitat.—Most typical on cool, moist sites.

Distribution.—Maine to Michigan, south to northeastern Alabama and northern Georgia (not on the coast below Delaware).

Remarks.—Sweet birch and yellow birch are the only birches from which oil of wintergreen can be

Fig. 57.—Young black birch, with yellow birch in the background.

obtained by distilling the twigs and inner bark. Of the two, sweet birch will yield by far the most, so that yellow birch is probably not treated for this purpose.

Birch beer is made from the sap of sweet birch. Gibson says to tap the tree, put the sap together with a handful of corn in a jug, and let fermentation do the rest!

Josselyn wrote of the inner bark of black birch and white birch, "the bark is used by the Indians for bruised wounds and cuts, boyled very tender and stampt betwixt two stones to a plaister, and the decoction thereof poured into the wound; and also to fetch Fire out of burns and scalds."

As a tonic, Seton suggests boiling 2 lb. of twigs to a gallon of water, until a pint of strong brown tea is left, which can be sipped a half pint a day.[1]

Paper Birch   White Birch

(*Betula papyrifera* Marsh.)

Appearance.—A medium-sized tree, 50 to 70 ft. high and 1 to 2 ft. in diameter (max. 120 by 4 ft. in the west), with a long trunk and open, pyramidal crown which later becomes rounded. Trees often occur two or more in a group.

Leaves.—Alternate (on spur shoots appearing opposite or whorled) simple, 2 to 3 in. long, 1½ to 2 in. wide, oval, doubly serrate, smooth above, more or less sticky below.

Flowers.—Male and female borne in separate catkins on the same tree; preformed male catkins present during the winter.

Fruit.—A minute winged nutlet (seed) borne in large numbers in a cylindrical cone whose scales fall readily when mature; cone, pendent, about 1½ in. long.

Twigs.—Reddish brown, slender; terminal bud, lacking except on spur shoots which are conspicuous features: lateral buds, broadest near the base, sticky when pressed between the fingers.

Bark.—At first, and usually until the tree is several inches in diameter, reddish brown, then peels off

[1]Birch tea is best made, however, by steeping, not boiling, the twigs. Boiling tends to drive off the wintergreen oil.

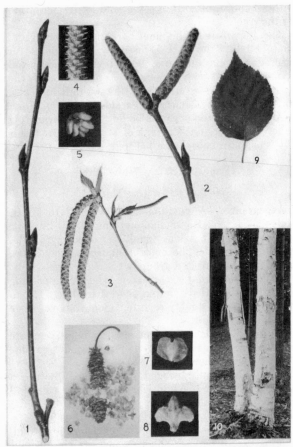

Fig. 58.—Paper birch. *1.* Twig × 1¼. *2.* Preformed male catkins × 1¼. *3.* Flowers; male catkins pendent, the female one, lateral × ½. *4.* Portion of female catkin × 5. *5.* Male flower × 5. *6.* Fruiting cone shedding its scales and winged nutlets. × ¾. *7.* Fruit × 3. *8.* Bract, or cone scale × 3. *9.* Leaf × ½. *10.* Bark of young tree.

to show the new white layers beneath; eventually, entirely chalky white and peels off in papery curls; on old trees, black near the base of the trunk.

Habitat.—Found mostly on moist sandy soils; especially common after fires, mixed with trembling aspen and fire cherry.

Distribution.—For the most part, a Canadian species ranging from Labrador to Alaska; in the

WHITE BIRCH BARK
THE PARCHMENT OF THE FOREST
Write a letter home on it, use it to make a folding drinking cup; or a dipper for the spring. Water can be boiled and tea made in a birch bark kettle. The Canadian Indians still use it for canoes. When gathering birch bark—

"NEVER PEEL A STANDING TREE"

Either pick up from the ground naturally shed bark or find where several trees are growing in a clump away from the trail or road. Cut down the poorer ones and take their bark. This helps the remaining trees by giving them more light.

Fig. 59. ——Paper birch bark.

United States, found in New York, New England, and as far south as northern Pennsylvania, passing westward just below the Great Lakes and through Iowa northward to Wyoming and into Canada (also scattered areas in the mountains of North Carolina and probably Tennessee).

Remarks.—Much has been written about the uses of paper birch bark; its most classical use was, and in some localities still is, for the Indian's canoe. It also has a host of other uses: as tinder for starting a fire, for bark utensils of all sorts, for "writing paper" (pen or typewriter), and as covering for lean-tos and tepees, to mention but a few. The pioneers put sheets of bark under their shingles. Bark should *never be peeled* from a standing tree. Naturally shed sheets can often be found; and where trees are plentiful, one tree from a group can be cut down and the bark removed. Peeled trees do not develop new white bark, but instead a black ring remains where the bark was removed. The wood is easily worked and was used by the northern Indians for snowshoe frames and paddles. For emergency food the inner bark is pounded to yield a flour, and syrup can be made from the sap, which itself is a pleasant cooling drink.

Gray Birch
(*Betula populifolia* Marsh.)

Appearance.—A small tree, 20 to 30 ft. high and 6 to 12 in. in diameter (max. 50 by 1½ ft.), with a

Fig. 60.—Gray birch. *1*. Leaf × ½. *2*. Bract, or cone scale × 3. *3*. Fruit × 3. *4*. Fruiting cone × ¾. *5*. Lateral bud × 1¼.

slender trunk and open, pyramidal crown. A common feature is the occurrence of several trees in a group, all apparently coming from the same root system.

Leaves.—Alternate (on spur shoots appearing opposite or whorled), simple, 2½ to 3 in. long, 1½ to

2½ in. wide, triangular, narrowly pointed, and doubly serrate, essentially smooth above and below, somewhat sticky when young.

Flowers.—Male and female borne in separate catkins on the same tree; preformed male catkins present during the winter.

Fruit.—A minute winged nutlet (seed) borne in large numbers in a cylindrical cone whose scales fall readily when mature; cone, pendent or spreading, about ¾ in. long.

Twigs.—Reddish brown, slender, covered with numerous warty glands; terminal bud, lacking except on spur shoots which are conspicuous features; lateral buds, spindle shaped, widest near the middle, somewhat sticky.

Bark.—At first dark brown, later dull grayish white, smooth, does not peel so readily as that of white birch.

Habitat.—Common and prolific, even on the poorest of dry, sterile soils.

Distribution.—Newfoundland south to northern Delaware, southwest along the St. Lawrence River and to western New York and Pennsylvania (isolated areas in Indiana at the foot of Lake Michigan).

Remarks.—Gray birch is one of the most characteristic and aggressive trees of New England, commonly in mixture with pitch pine and scrub oak. On better soils, gray birch is also found with young white pine; and although at first it serves as a protection, later it crowds out the pine. This birch follows fire much as does trembling aspen and, like it, is short-lived and more or less of a "weed tree," although the wood is used for spools and other small articles.

River Birch   Red Birch
(*Betula nigra* L.)

River birch is essentially a southern tree and the only birch at low elevations in the south. The leaves are rhombic-ovate and deeply doubly serrate; the cone scales and nutlets are hairy and, in contrast to those of the other native birches, mature in the late spring; salmon-pink may be used to describe the color of the papery bark which on old trees becomes reddish brown and roughened. This tree, as its

FIG. 61.—River birch. *1*. Fruiting cone × ¾. *2*. Bract, or cone scale × 3. *3*. Fruit × 3. *4*. Leaf × ½. *5*. Bark.

name indicates, is found along stream banks; and although rare in the north, it is common from Pennsylvania southward. River birch is distributed from southern New Hampshire through Pennsylvania to southeastern Minnesota, south to eastern Texas and northern Florida.

European White Birch

(*Betula pendula* Roth.)

European white birch was introduced as an ornamental and is now found throughout the eastern

United States. The leaves resemble those of gray birch but are more rounded at the base and not so long pointed; the bark is similar to that of the native white birch but peels less. In some localities, the European form has not proved satisfactory because of its susceptibility to the attacks of the bronze birch borer whose larvae riddle the trunk. According to Sargent the bark contains a resin (betulin) that accounts for its durability. "Russian leather" has an odor from the birch oil and tannin used in its preparation. The bark has also been used for the soles of shoes.

FIG. 62.—Leaf of European white birch × ½.

THE ALDERS

Speckled Alder

[*Alnus rugosa* (DuRoi) Spreng.]

Speckled alder is usually a large shrub but sometimes may attain a height of 20 ft. and a diameter of 4 or 5 in. It is exceedingly common along northern streams or in swamps. The leaves are alternate, 2 to 4 in. long, broadly elliptical to oval, and coarsely, doubly serrate almost to the point of being lobed. The woody cones are persistent until the following season and may usually be seen on the twigs. Perhaps the best feature for recognizing this species is the plainly visible triangular pith (cross section). The birches may also have a triangular pith, but it is so small that a pocket lens is necessary to detect its shape.

This alder growing in such abundance along streams serves to protect the banks from excessive erosion and also furnishes cover for small game birds. It ranges from Newfoundland to northwestern Canada, south to Pennsylvania and Nebraska.

Smooth alder [*A. rugosa* (Du Roi) Spreng.] may be recognized by its leaves, which are broader near the tip and acute at the base, with evenly serrate margins. Seton says that a tea from the leaves is a valuable tonic and skin wash for pimples.

The bark and cones of the alders contain tannin, and the Eskimos dye reindeer skins with bark extracts. Seton states that an orange dye is thus produced by boiling with the inner bark.

Alders are useful to wild life in a number of ways. Bees use the pollen for spring-brood rearing; such birds as the sharp-tailed and ruffed grouse eat the leaves and buds (also to some extent the seeds); and rabbits and beaver chew the bark.

Hophornbeam   Ironwood

[*Ostrya virginiana* (Mill.) K. Koch.]

Fig. 63.—Speckled alder.
1. Leaf × ½. *2.* Buds × 1.
3. Fruiting cones × ¾.

Appearance.—A small tree, 20 to 35 ft. high and 6 to 15 in. in diameter (max. height 70 ft.), with a large, rounded crown and cylindrical trunk.

Leaves.—Alternate, simple, 2½ to 5 in. long, 1½ to 2 in. wide, elliptical to ovate, with finely doubly serrate margins, smooth above; smooth below except for tufts of hair in the vein axils.

Flowers.—Borne separately on the same tree; the male in catkins, the female in spikes or short catkins; preformed male catkins present during the winter.

Fig. 64.—Hophornbeam. *1.* Leaf × ½. *2.* Fruit × ¾. *3.* Fruit with part of envelope removed showing seed × ¾. *4.* Bud × 1. *5.* Bark.

Fruit.—A nutlet, borne in a bladdery sac, a number of which form a conelike cluster resembling a hop.

Twigs.—Brownish, slender, tough; terminal bud, lacking; laterals, ovoid, circular in cross section.

Bark.—Brownish, soon shredding off in narrow scaly plates which curl at the free ends.

Habitat.—A common understory species in hardwood forests, a weed tree with the ability of growing in dense shade.

Distribution.—From Nova Scotia to Manitoba, south to eastern Texas and Florida.

Remarks.—The wood is extremely hard and tough, hence the name *ironwood*. Because of the small size of the tree, it is of little use except for firewood (hard to work) and for such items as bows, handles, wedges, and sled runners. The bark, although rich in tannin (Sargent), is little used.

The buds are important winter food of such birds as the ruffed grouse, bobwhite, and ptarmigan.

American Hornbeam   Blue-beech   Water-beech

(*Carpinus caroliniana* Walt.)

Appearance.—A small tree (often shrubby), 20 to 30 feet high and 4 to 8 in. in diameter (max. 40 by 2 ft.), with a short, irregular, fluted trunk and round-topped crown.

Leaves.—Alternate, simple, 2 to 4 in. long, 1½ to 2 in. wide, oblong-ovate, finely doubly serrate, thin and firm in texture, smooth above, smooth or finely hairy below.

Flowers.—Borne separately on the same tree; the male in catkins, the female in spikes or short catkins. (This is the only native species in the birch family in which the partially developed, preformed male catkins are not found on the tree during the winter.)

Fruit.—A nutlet borne at the base of a three-lobed leafy bract, a number of which are arranged spirally to form a conelike cluster.

Twigs.—Dark reddish brown, slender, tough; terminal bud, lacking; the laterals, of two sorts, (1) large (flower buds), (2) small and more or less four-angled in cross section (leaf buds).

Bark.—Smooth, dark bluish gray.

Habitat.—Found mostly as an understory weed tree in hardwood forests on moist or wet soils or along stream banks.

Distribution.—Nova Scotia to central Minnesota, south to eastern Texas and Florida.

Remarks.—Unfortunately, this tree has acquired the name *beech* (water or blue) even though it is not in the beech but rather in the birch family. A better name seems to be American hornbeam, although usage will probably keep alive the present misnomer. The wood is hard and heavy and suitable for the same items listed under the previous species. The buds are also eaten by several birds including ruffed grouse and bobwhite.

THE BEECH FAMILY

American Beech

(*Fagus grandifolia* Ehrh.)

Appearance.—A medium-sized tree, 70 to 80 ft. high and 2 to 3 ft. in diameter (max. 120 by 4 ft.), with a short trunk and wide-spreading crown. Sprouts from

Fig. 65.—Leaf ×
½, and fruit × ¾ of
American hornbeam.

the shallow root system often form thickets around old trees.

Leaves.—Alternate, simple, 2½ to 5 in. long, 1 to 2½ in. wide, elliptical, with a serrate margin, each principal vein ending in a tooth, smooth above,

Fig. 66.—Beech. *1*. Twig × 1¼. *2*. Flowers × ½. *3*. Leaf
× ½. *4*. Fruit × ¾. *5*. Fruit after opening showing triangular
nuts × ¾. *6*. Bark.

smooth below except for hair in the vein axils; texture of the leaf, thin and papery to the touch.

Flowers.—Borne separately on the same tree; the male in globose heads suspended by long stems, the female in short spikes.

Fruit.—A characteristic nut, $\frac{1}{2}$ to $\frac{3}{4}$ in. long, triangular in cross section and enclosed usually in pairs in a four-parted bur covered with weak, unbranched spines; nut, sweet, edible.

Twigs.—Brownish gray, slender, more or less zigzag; terminal bud, present, similar to the laterals, $\frac{3}{4}$ in. long and lance- or cigar-shaped with many scales.

Bark.—Throughout life, smooth, blue-gray, sometimes mottled or warty.

Habitat.—Found on a wide variety of soils, but nearly always where abundant surface moisture is available.

Distribution.—New Brunswick to eastern Wisconsin, south to eastern Texas and northern Florida.

Remarks.—Beech together with yellow birch and sugar maple form the so-called northern *beech-birch-maple* forest which covers wide areas in the northeast and Lake states. However, beech reaches greatest size and best development farther south and, during the exceptionally cold winter of 1934, suffered considerable damage in the northeast. Together with sugar maple, it can grow under heavier forest cover than other associated hardwoods, and its tough branches seem to take advantage of newly cleared trails to develop more rapidly than those of other species. The wood is used for many purposes including fuel for which it rates highly. The nuts are important as food for many small animals including mice and squirrels and are also eaten by bear, raccoon, turkey, and ruffed grouse. Because of its smooth gray bark, beech usually has a great attraction for that species of forest park vandal, the initial carver. Nothing good can be said for the practice of trying to attain a cheap immortality by scarring an otherwise attractive tree trunk.

Oil pressed from the nuts of the European beech has been used to adulterate olive oil, for making soap, and for illuminating purposes. Since the native beech is so similar, its nut could probably be used in the same way. The leaves of the European beech were used to stuff mattresses and were preferred to straw because they lasted for a number of years without getting musty. It was natural that this custom should be brought to this country where in the Ohio Valley, at least, the native beech was similarly used. The leaves are more springy than those from other trees and do not mat down so readily. The wood ashes contain appreciable amounts of potash and were important in soapmaking.

It has been frequently said that beech trees are never struck by lightning; the author will much appreciate hearing from anyone to the contrary.

American Chestnut[1]

[*Castanea dentata* (Marsh.) Borkh.]

Appearance.—A medium-sized to large tree, 70 to 90 ft. high and 3 to 4 ft. in diameter (max. 120 by 10 ft.).

Leaves.—Alternate, simple, 6 to 8 in. long, about 2 in. wide, oblong-lance-shaped; sharply coarsely serrate with each principal vein ending in a hairlike point, smooth above and below.

Flowers.—Borne in catkins on the same tree; catkins of two kinds (1) long and bears male flowers

---

[1]Chinkapins are small trees of the chestnut group with smaller fruit and leaves whitish woolly beneath, ranging from the central states southward.

Fig. 67.—Chestnut. *1*. Twig × 1¼. *2*. Leaf × ½. *3*. Flowers × ⅓. *4*. Portion of male catkin showing clusters of flowers × 1. *5*. Basal portion of bisexual ament showing a female flower × 1. *6*. Nut × 1. *7*. Fruit × ¾. *8*. Chestnut sprouts (old trunks blight-killed).

only, (2) short and bears both female and male flowers.

Fruit.—A rounded nut about ¾ in. long, borne in 2s or 3s in a bur covered with long, needle-sharp, branched spines; nut, sweet, edible.

Twigs.—Reddish brown, slender to moderately stout, with star-shaped pith; terminal bud, lacking; laterals, ovoid with two or three visible scales; leaf scars, half round.

Bark.—Dark brown, shallowly furrowed, with broad flat ridges.

Habitat.—Found on many soil types, but probably makes best growth on moist sandy loams.

Distribution.—Southern Maine to southern Michigan, south to northern Mississippi and northwestern Florida (below Virginia, not on the coastal plain).

Remarks.—During the last century, chestnut was one of the most important eastern hardwoods. The tree was fast growing; stump sprouts quickly produced new trees after the old growth was logged. The nuts were eaten in large numbers, and the soft, brown wood was durable in contact with the soil. In the early nineteen hundreds, a fungous disease on chestnut reached the U. S. from eastern Asia. The Asiatic chestnuts were only mildly susceptible, having developed an immunity over many centuries, but the fungus proved to be deadly to our native trees. Large sums were spent in an effort to find a point of attack somewhere in its life history but with no success. The blight, which kills the inner living bark, has now spread over most of the range of chestnut, and few if any of the original trees remain. The remarkable vitality of this species has resulted in numerous sprouts around the old dead trunks. These sprouts often attain a height of 20 ft. or more before succumbing to the fungus which continues to live in portions of the roots. These sprouts often bear nuts, some of which will grow and produce seedlings. The Brooklyn Botanical Garden is planting nuts sent by various interested people, in the hope that somewhere may be found a single tree immune to the disease. In most if not all cases, the trees from nuts collected on sprouts are no more resistant than were the old trees from which they sprang. This botanical garden and other agencies are also experimenting with hybrids in the hope of obtaining resistant strains. The wiping out of an entire tree population like this means that strict plant quarantine laws must be enforced to prevent a similar calamity to others of our forest trees (see American elm).

Fig. 68.—Oak. *1.* Flowering branchlet × ½. *2.* Female flowers × 4. *3.* Male flower × 4.

Strachey, an early writer, said "In some places we fynd chestnuts whose wild fruict I maie well saie equallize the best in France, Spaine, Germany, Italy, or those so commended in the Black Sea, by Constantinople of all which I have eaten."

According to Seton, "the nut of this tree is hung high aloft wrapped in a silk wrapper which is enclosed in a case of sole leather, which again is packed in a mass of shock absorbing vermin-proof pulp,[1] sealed up in a water-proof iron-wood safe and finally cased in a vegetable porcupine of spines, almost impregnable."

---

[1] Heavy velvet might be better.

THE OAKS

Leaves.—Alternate, simple, usually lobed, with or without bristle tips; extremely variable in shape.

Flowers.—Male and female borne on the same tree; the former in pendent stringy catkins, the latter in small inconspicuous clusters.

Fruit.—An acorn with a scaly cup which varies from saucer-shaped to bowl-like depending upon the species.

Fig. 69.—Variation in the shape of post-oak leaves all from the same tree.

Twigs.—Slender to moderately stout; buds, alternate; terminal, present, many scaled; pith, star-shaped in cross section.

Remarks.—The oaks as a group are among our most important and common broadleaved or hardwood trees. The wood is hard and strong, suitable for many purposes. Used as firewood, it burns to a hot bed of coals suitable for broiling. The bark of most oaks contains tannin, and that of some, dye substances, extractable with hot water. The leaves and twigs are noted for their display of galls, swellings caused by various wasplike insects which lay their eggs in the living tissues. When one cuts into one of these galls, a small grub is usually found near its center. Some galls are especially high in tannin, from which ink can be made by mixing the water extract with a solution of an iron (ferric) salt.

Deer eat the acorns and browse the twigs in winter; the former is also true of such large birds as the turkey, grouse, and pheasant.

┌─ OAKS ─┐
WHITE   RED

	White	Red
Leaves:	With rounded lobes or teeth	With bristle- or hair-tipped lobes, or leaves
Fruit:	Maturing in one season	Maturing in two seasons
Seed:	Relatively sweet and edible	Bitter, not edible

KEY TO THE OAKS[1]

1. Leaf tips or lobes, without hairs or bristles; inner surface of nut shell, smooth; seed, relatively sweet, edible (*White oaks*). **2**
1. Leaves, with fine hair or bristle tips at the ends; inner surface of nut shell, lined with wool; seed, bitter (*Red* or *Black oaks*). **9**

2. Leaves, serrate, or with rounded or scalloped teeth (*Chestnut oaks*). **3**
2. Leaves, cut into segments (lobes) by distinct gulfs or sinuses. **5**

[1] Since oaks are so variable, it is possible to collect leaves that cannot be properly identified by this or any other key to "normal" material.

3. Teeth, "sharp," but the very ends slightly rounded like a nipple......... **Chinkapin oaks** (p. 153)
3. Teeth, coarsely rounded, or scalloped............. **4**

    4. Margin, irregular, some sinuses much deeper than others, often almost lobed, brownish velvety below..........**Swamp white oak** (p. 151)
    4. Margin, regularly wavy or with rounded coarse teeth, smooth or inconspicuously hairy below... **Chestnut oak** (p. 150)

5. Leaves, shaped like a cross.......**Post oak** (p. 148)
5. Leaves, not cross-shaped........................ **6**

    6. Leaves with a center pair of sinuses going nearly to the midrib; nut, one-half or more enclosed in a deep, fringed cup...........**Bur oak** (p. 146)
    6. Leaves, with sinuses all deep or all shallow, not as above; nut, enclosed only about one-third by the cup................................... **7**

7. Sinuses, reaching nearly to the midrib............ **White oak** (p. 114)
7. Sinuses, reaching not more than halfway to the midrib.. **8**

    8. Leaves, regularly lobed; smooth below......... **Northern white oak** (p. 146)
    8. Leaves, irregularly lobed, velvety below....... **Swamp white oak** (p. 151)
9. Leaves, willowlike, unlobed (*Willow oaks*)......... **10**
9. Leaves, lobed................................. **11**

    10. Leaves, narrow, shiny, smooth.............. **Willow oak** (p. 167)
    10. Leaves, elliptical, hairy below and on the leaf stem....................**Shingle oak** (p. 165)

11. Leaves, mostly 3-lobed, broad at the tip and tapering downward, triangular....... **Blackjack oak** (p. 163)
11. Leaves, usually 5- or more lobed, elliptical to circular in outline................................. **12**

    12. Leaves, small, about 3″ long, mostly narrow, more or less oblong........ **Scrub oak** (p. 165)
    12. Leaves, 4″ or more long, oval to circular....... **13**

13. Sinuses, deep, nearly to the midrib............... **14**
13. Sinuses, extending about halfway to the center..... **17**

    14. Sinuses, nearly circular in outline; meat of nut, white; concentric lines, often present around tip of acorn.................. **Scarlet oak** (p. 159)
    14. Sinuses, squarish or opening outward; meat of nut, yellowish; circles, lacking............... **15**

15. Leaves, more or less scurfy (like fine dandruff) below (rub with the thumb)........... **Black oak** (p. 157)
15. Leaves, smooth, often shiny below.............. **16**

    16. Acorn, short and wide; nut, enclosed only at the base by a flat saucerlike cup... **Pin oak** (p. 161)
    16. Acorn, narrow, ellipsoidal; nut, partly enclosed by a deep cup.................................
        Northern pin oak, Hill's oak (p. 163)

17. Lobes, 7 to 11; sinuses tend to be of about the same depth; acorn, $\frac{3}{4}$″ long (*Red oak*)................. **18**
17. Lobes, 5 to 7, more or less scurfy (like dandruff) below; sinuses, irregular; acorn, about $\frac{2}{3}$″ long....
        Black oak (p. 157)

    18. Acorn, ellipsoidal; nut, partly enclosed by a deep cup........... **Northern red oak** (p. 155)
    18. Acorn, stout; nut, enclosed only at the base in a thick saucerlike cup.. **Common red oak** (p. 155)

White Oak

(*Quercus alba* L.)

Appearance.—A medium-sized to large tree, 80 to 100 ft. high and 3 to 4 ft. in diameter (max. 150 by 8 ft.), with a short, stocky trunk and wide-spreading, rounded crown.

Leaves.—Alternate, simple, 5 to 9 in. long, 2 to 4 in. wide, oval in outline or broader near the apex, smooth above and below; deeply lobed; margins of lobes, entire.

Flowers.—Borne separately on the same tree; the male in catkins, the female in short spikes.

Fruit.—An acorn, about ¾ in. long; the cup, covered with warty scales and enclosing the nut about one-fourth of its length; kernel, sweet.

Twigs.—Grayish or greenish red, often with a purplish tinge, moderately stout; pith, star-shaped; terminal bud, present, nearly globose, accompanied by two or three lateral buds clustered at the tip; bud scales, many.

Bark.—Light ashy gray, variable in texture, scaly or divided into small rectangular blocks.

Habitat.—Although found on many soil types, best growth is made on deep, rich, well-drained soils, such as in the higher bottom lands.

Distribution.—From southern Maine to southeastern Minnesota, south to eastern Texas and northern Florida.

Remarks.—This tree together with its variety (see below) is the most common and important of eastern white oaks and contributes about three-fourths of the wood sold under this name. As with most of the other white oaks, the acorns of this species start to sprout soon after they fall in autumn, and many of them are frozen before the roots can penetrate the ground. They are a source of food for the

FIG. 70.—White oak. *1*. Twig × 1¼. *2*. Leaf of true white oak × ½. *3*. Leaf of "northern white oak" × ½. *4*. Acorn × 1. *5*. Bark. *6*. Bark of very old tree.

gray squirrel which "accidentally" plants many of them at a time when they must be covered or die. That this is important in establishing future oak forests has been pointed out by Seton who suggests that a diminishing squirrel population means fewer white oaks. The acorns were also eaten by the Indians and settlers, who made the acorns more palatable by boiling them in water. According to an early writer, "They have in Virginia a goodly Oke which they call the white Oke because the barke is whiter, than of others,—the Ackorne likewise, is not onely sweeter than others, but by boyling it long it giveth an oyle which they keepe to supple their joynts."

The wood is one of our best hardwoods and is used for many purposes including furniture, flooring, barrel staves, railroad ties, and fuel. (In colonial times, it had first place in shipbuilding.) Michaux, in 1857, mentioned that white oak was used in Maine for axe handles. This custom still persists presumably because of a scarcity of hickory there but perhaps on account of choice, since the "feel" of the two woods is not identical. In colonial days, tanbark was obtained from this tree and also considerable charcoal.

Northern White Oak

(*Quercus alba* var. *latiloba* Sarg.)

Since both shallowly and deeply lobed leaves can often be found on the same tree, this variety is no longer recognized. It is considered synonymous with white oak.

Bur Oak　　Mossy Cup Oak

(*Quercus macrocarpa* Michx.)

Appearance.—A medium-sized to large tree, 70 to 80 ft. high and 2 to 3 ft. in diameter (max. 170 by

7 ft.), with a short, massive trunk and large, rounded crown.

Fig. 71.—Bur oak. *1.* Twig × 1¼. *2.* Leaf × ½. *3.* Acorn × ¾. *4.* Bark.

Leaves.—Alternate, simple, 6 to 10 in. long, 2 to 4 in. wide, oval or broadest near the tip, deeply and irregularly lobed, dark green and smooth above, paler and hairy below. The center pair of sinuses (spaces between lobes) usually reach nearly to the midrib; lobe margins, entire.

Flowers.—Borne separately on the same tree; the male in catkins, the female in short spikes.

Fruit.—An acorn, ¾ to 1½ in. long (max. 2 in. in the south). The cup, with a conspicuously fringed edge, covers one-half or more of the nut; kernel, sweet.

Twigs.—Yellowish brown, stout, usually corky after the first year; pith, star-shaped; terminal bud, present, short, usually accompanied by two or three clustered lateral buds; bud scales, many.

Bark.—Dark gray, furrowed into roughly rectangular plates or coarsely scaly.

Habitat.—A typical wet bottom-land tree; but when planted, prospering on dry soils and found naturally upon them in the west.

Distribution.—Nova Scotia to Manitoba, south to central Texas, thence northeast to Delaware.

Remarks.—Bur oak shows perhaps the greatest diversity in size of any eastern white oak. In the Wabash River Valley, it is said to attain the maximum height of 170 ft.; but on the foothills of the Rockies, it is reduced to a small tree or shrub. Its peculiarly lobed leaves, fringed acorns, and corky twigs make it a recommended ornamental, as well as the fact that it seems to withstand smoke and gas better than certain other species.

Post Oak

(*Quercus stellata* Wang.)

Appearance.—A small to medium-sized tree, 40 to 50 ft. high and 1 to 2 ft. in diameter (max 100 by 3½ ft.), with a short trunk and large, rounded crown.

Leaves.—Alternate, simple, 4 to 6 in. long, 3 to 4 in. wide, oblong-oval or broadest above the middle, lobed in such a way that the leaf suggests a cross in shape; margin of lobes, entire, with tufts of hair above, yellowish woolly below.

Flowers.—Borne separately on the same tree; the male in catkins, the female in short spikes.

Fruit.—An acorn, about ⅔ in. long; the bowl-shaped cup (with scales less warty than those of white oak) enclosing about one-third of the nut.

Fig. 72.—Post oak. *1.* Twig × 1¼. *2.* Acorn × 1. *3.* Leaf × ½. *4.* Bark.

Twigs.—Tawny woolly, with star-shaped pith; terminal bud, present, nearly globose, accompanied by two or three laterals clustered at the tip; bud scales, many.

Bark.—Reddish brown, with narrow, vertical, more or less rectangular ridges.

Habitat.—Found mostly on poor, dry soils not able to support some of the better hardwoods.

Distribution.—From Massachusetts to southern Iowa, south to central Texas and northern Florida.

Remarks.—Not a common tree in the northeast and often reduced to a shrub at its northern limits.

Chestnut Oak

(*Quercus prinus* L.)

Appearance.—A small to medium-sized tree, 40 to 60 ft. high and 2 to 3 ft. in diameter (max. 100 by

FIG. 73.—Chestnut oak. *1.* Twig × 1¼. *2.* Leaf × ½. *3.* Acorn cup showing thin flat sides × 1. *4.* Acorn × 1. *5.* Bark. (*Photograph by H. P. Brown.*)

7 ft.), with a short, stocky trunk and broadly rounded crown.

Leaves.—Alternate, simple, 4 to 8 in. long, 1½ to 3 in. wide, elliptical or widest above the middle, coarsely toothed with rounded teeth; smooth above, paler and often finely hairy below.

Flowers.—Borne separately on the same tree; the male in catkins, the female in short spikes.

Fruit.—An acorn about 1 in. long, the nut enclosed one-third of its length in a thin cup whose inner surface slopes downward with little curvature (flat sided).

Twigs.—Orange-brown, moderately stout, with star-shaped pith; terminal bud, present, $\frac{1}{4}$ in. long, acute, accompanied by two or three laterals clustered at the tip; bud scales, many.

Bark.—Thick, deeply and heavily furrowed.

Habitat.—Found mostly on poor, dry soils because of its inability to compete on better sites with other hardwoods.

Distribution.—For the most part, an Appalachian and Ohio Valley tree; from southern Maine to southern Illinois, south to the District of Columbia and along the mountains to northern Georgia (also scattered areas in southern Ontario and Michigan).

Remarks.—Chestnut oak prevents erosion on poor sites often barren of other tree growth. On better soils it reaches commercial proportions and is used as already indicated under white oak. The bark was considered the best source of white oak tannin, which was combined with that from hemlock to offset the red color of the latter.

Swamp White Oak

(*Quercus bicolor* Willd.)

Appearance.—A medium-sized tree, 60 to 70 ft. high and 2 to 3 ft. in diameter (max. 100 by 7 ft.), with a short trunk and rounded crown of large branches whose bark tends to be shed in ragged papery flakes.

Leaves.—Alternate, simple, 5 to 6 in. long, 2 to 4 in. wide, widest above the middle, smooth above, minutely woolly below; margin, irregular, may be coarsely toothed to shallowly lobed.

Flowers.—Borne separately on the same tree; the male in catkins, the female in short spikes.

Fruit.—An acorn, about 1 in. long, the nut enclosed for one-third of its length in a slightly fringed, bowl-shaped cup, usually borne on a slender stem 1 to 4 in. long.

Fig. 74.—Swamp white oak. *1.* Buds × 1¼. *2.* Leaf × ½. *3.* Acorn × 1. *4.* Bark.

Twigs.—Straw-brown, slender or moderately stout; pith, star-shaped; terminal bud, present, short, rounded, and accompanied by two or three clustered laterals; bud scales, many.

Bark.—At first ragged, later becomes heavily furrowed and with long flat ridges.

Habitat.—As indicated by its common name, a swamp and bottom-land tree.

Distribution.—From southern Maine to southern Minnesota, south to Arkansas; along the Atlantic Coast to Delaware and in the mountains to South Carolina.

Remarks.—Not a common tree but, when present, may be abundant locally.

Chinkapin Oak

(*Quercus muehlenbergii* Engelm.)

Appearance.—In the northeast, a small to medium-sized tree 20 to 50 ft. high and 1 to 2 ft. in diameter.

FIG. 75. ——Chinkapin oak. *1.* Twig × 1¼. *2.* Acorn × 1. *3.* Leaf × ½. *4.* Bark.

Farther west in the Ohio Valley, the maximum size of 160 by 4 ft. may be attained. The trunk is short and often buttressed, and bears a rounded crown.

Leaves.—Alternate, simple, 4 to 7 in. long, 1 to 4 in. wide, oblong-elliptical, broadest above the middle, or nearly lance-shaped; with coarse teeth inconspicuously rounded at the tips; smooth above, pale and finely hairy below.

Flowers.—Male and female borne separately on the same tree; the male in catkins, the female in short spikes.

Fruit.—An acorn, about ¾ in. long, with a thin, bowl-shaped cup enclosing one-third to one-half of the nut.

Twigs.—Orange-brown, slender, with star-shaped pith; terminal bud, ⅛ in. long, pointed, accompanied by two or three clustered laterals.

Bark.—Ashy gray, more or less rough and flaky.

Habitat.—In the northeast found commonly on limestone outcrops, farther west in bottom lands.

Distribution.—From southern Vermont to northern Iowa, south to central Texas and northern Florida (rare or lacking on the coastal plain below Maryland).

Remarks—Chinkapin oak is of little importance in the northern states because of its usual small size and relative scarcity. Michaux said that the acorns are sweeter than those of any other oak in the United States. Experience shows that they *are* quite "sweet" and palatable.

Dwarf Chinkapin Oak

(*Quercus prinoides* Willd.)

Usually shrubby, this species occasionally reaches tree size. It is similar to chinkapin oak except that (1) the leaves are smaller with shorter leaf stems and more wavy margins, (2) the nut is enclosed more deeply by the cup, (3) the cup scales are somewhat warty. Found on poor dry soils, it ranges from Massachusetts to Minnesota, south to eastern Texas and North Carolina.

Northern Red Oak

(Quercus rubra L.)

Appearance.—A medium-sized tree, 60 to 70 ft. high and 2 to 3 ft. in diameter (max. 150 by 5 ft.), with a comparatively short massive trunk and rounded crown of large branches.

Leaves.—Alternate, simple, 5 to 8 in. long, 4 to 5 in. wide, oblong to oval or widest above the middle, seven- to eleven-lobed, each lobe forked and bristle tipped; smooth above, smooth below except for occasional tufts of hair in the vein axils.

Flowers.—Borne separately on the same tree; the male in catkins, the female in short spikes.

Fruit—An acorn usually about 1 in. long, the nut enclosed only at the base in a thick, saucer-shaped cup; kernel; bitter. (As Seton aptly remarks, red oak, strictly speaking, "has no cup, it has a saucer.")

Twigs.—Reddish brown, moderately stout; pith, star-shaped; terminal bud, $\frac{1}{4}$ in. long, accompanied by two or three clustered laterals; bud scales, many, more or less hairy.

Bark.—Dark brown, with broad, flat ridges; on very old trees heavily corrugated.

Habitat.—Found on many different sites but does best on rich, moist, well-drained soils.

Distribution—From Nova Scotia to eastern Minnesota, south to Arkansas and Northern Georgia (not on the coastal plain below Chesapeake Bay).

Remarks—Formerly, trees with flat-cupped acorns were called "common red oak," those with narrow, deep-cupped acorns, "northern red oak." It has been found that the two forms do not breed true and that there are many intermediate acorn shapes. Therefore it seems best to consider this red oak as a single species with variable acorns. It grows farther north than any

For descriptive legend see opposite page.

of the other red oaks and has proved the most hardy of American oaks grown in Europe. The wood is inferior to that of the white oaks but is used for similar purposes[1] (see white oak). Red oak makes rapid growth and is used to some extent as ornamental.

Black Oak

(Quercus velutina Lamarck)*

Appearance.—A medium-sized tree, 50 to 60 ft. high and 2 to 3 ft. in diameter (max. 150 by more than 4 ft.), with a short trunk and somewhat oblong crown.

Leaves.—Alternate, simple, 5 to 7 in. long, 3 to 5 in. wide, nearly oval in outline, ovate or broadest above the middle, with five or seven forked and bristle-tipped lobes; smooth, shiny, and dark green above; yellowish green, more or less hairy below and scurfy (like dandruff).

Flowers.—Borne separately on the same tree; the male in catkins, the female in short spikes.

Fruit.—An acorn, about ¾ in. long; the nut, enclosed one-third in a bowl-shaped cup with chestnut-brown, dull scales; seed, yellowish, bitter.

Twigs.—Reddish brown, moderately stout, with star-shaped pith; terminal bud, angled, accompanied by two or three clustered laterals; bud scales, grayish woolly.

Bark.—Thick, nearly black, heavily furrowed vertically, and with many, horizontal checks; inner bark, bright orange or yellowish.

[1] The large springwood pores in red oak are more open than in white oak; dip the end of a piece a foot long in soap solution, and blow on the other end!

FIG. 76.—Northern red oak. *1.* Twig × 1¼. *2.* Acorns at end of 1st season × 2. *3.* Narrow form of acorn × 1. *4.* Wide, more common form of acorn. × 1. *5.* Leaf × ½. *6.* Bark of young tree. *7.* Bark of old tree.

Fig. 77.—Black oak. *1.* Twig × 1¼. *2* and *3.* Leaves showing variation in depth of spaces between lobes (No. 2 more common) × ½. *4.* Acorns showing variation × 1. *5.* Bark.

Habitat.—Although makes best growth on good soils, is usually crowded out because of its inability to withstand competition; most frequent on poor, dry sites.

Distribution.—Southern Maine to Iowa, south to eastern Texas and northern Florida.

Remarks.—Black oak is probably more variable in its leaf and fruit characters than any other northern oak. Its leaves are often mistaken for those of red oak, and at times they are as deeply lobed as those of scarlet oak. The wood is inferior to that of red oak but is cut and used for the same purpose.

The bark contains tannin and also a yellow dye which were used in colonial times. Michaux says "From the cellular tissue of the black oak is obtained the quercitron, of which great use is made in dyeing wool, silk, and paper hanging." According to Gibson, the peeled bark was dried, pounded to powder, and sifted. The finer (yellow) particles pass through, leaving the coarser, dark-brown residue in the sieve; iron salts were used to modify the color.

Scarlet Oak

(*Quercus coccinea* Muench.)

Appearance.—A medium-sized tree, 70 to 80 ft. high and 2 to 3 ft. in diameter (max. 100 by 4 ft.), with a short trunk and irregular, rounded, or oblong crown.

Leaves.—Alternate, simple, 4 to 7 in. long, 3 to 5 in. wide, more or less oval in outline, with five to nine narrow, forked, bristle-tipped lobes separated by deep rounded sinuses (spaces); smooth and shiny above; smooth, except for tufts in the vein axils, below.

Flowers.—Borne separately on the same tree; the male in catkins, the female in short spikes.

Fruit.—An acorn about 1 in. long, the nut, enclosed one-third to one-half in a bowl-shaped cup with tightly appressed, somewhat shiny scales; nut, usually with one or more concentric circles around the tip; kernel, white, bitter.

Fig. 78.—Scarlet oak. *1.* Twig × 1¼. *2.* Leaf × ½. *3.* Acorn × 1. *4.* Top of acorn showing concentric rings × 3. *5.* Bark.

Twigs.—Reddish brown, slender, with star-shaped pith; terminal bud, present, often somewhat angled accompanied by two or three clustered laterals; scales, dark reddish brown with paler margins.

Bark.—Dark brown to nearly black, at first flaky, later roughly corrugated, similar in appearance to that of black oak, but not yellowish within.

Habitat.—Most common on dry, sterile, sandy soils.

Distribution.—Southern Maine to southeastern Minnesota (most of New York excepted), south to eastern Oklahoma and Georgia.

Remarks.—Quite often used as an ornamental tree because of its foliage, which turns scarlet in autumn.

Pin Oak

(*Quercus palustris* Muench.)

Appearance.—A medium-sized tree, 70 to 80 ft. high and 2 to 3 ft. in diameter (max. 120 by 5 ft.), with a short trunk and rounded or pyramidal crown. The trunk and branches display numerous short tough branchlets which, at a distance in winter, look like "pins" hence possibly the name. Also, the wood was used to make pins for holding together the squared timbers of barns.

Leaves.—Alternate, simple, 3 to 5 in. long, 2 to 5 in. wide, broadly oval in outline, with five or seven narrow, forked, bristle-tipped lobes separated by deep elliptical or angled sinuses (spaces) (compare with those of scarlet oak); smooth above, smooth below except for tufts of hair in the vein axils.

Flowers.—Borne separately on the same tree; the male in catkins, the female in short spikes.

Fruit.—An acorn, about $\frac{1}{2}$ in. long; the nut, enclosed only at the base in a thin saucerlike cup; kernel, bitter.

Twigs.—Reddish brown, slender, with star-shaped pith; terminal bud, present, accompanied by two or three clustered laterals, ovoid, usually not angled.

Bark.—Grayish brown, smooth for many years, eventually with low scaly ridges and shallow fissures.

Habitat.—A typical bottom-land tree reaching best development through the Ohio Valley.

Distribution.—Western Connecticut to Iowa (except most of New York), south to Arkansas and Virginia (along the Piedmont in North Carolina).

Remarks.—Although cut for lumber, the wood is knotty because of the many small branches found on the trunk. The tree is widely used as an ornamental on account of its pleasing appearance and ease of transplanting. (Most oaks are difficult to move because of their deep taproots.)

Fig. 79.—Pin oak. *1.* Twig × 1¼. *2.* Leaf × ½. *3.* Acorn × 1. *4.* Bark.

A German writer has said that each kind of tree has its own note or pitch produced when the wind blows through its crown. The pitch is the average of all the tones produced by the leaves on that tree. The idea that a tree can be recognized by listening to it may be new to some, but it is not so farfetched. Stand under a pin oak, and listen to the hiss of the wind as it is "torn to shreds" by the deeply lobed leaves and perhaps slender branches. Then stand near a poplar, and listen to the noisy clack of its many "tongues"; the two sounds are quite different. Of course the human ear is delicate enough to sense only such extremes as the two mentioned; probably there is little difference between pin oak and scarlet oak which have such similar leaves. But try this, and see how many trees you can tell by their "voices."

Gibson says that twig galls are especially common on pin oak. When soaked in a kettle of water containing a few nails, a black ink is produced. This ink will "eat up" a steel pen; hence it is necessary to use the primitive goose quill.

Northern Pin Oak   Jack Oak   Hills Oak

(*Quercus ellipsoidalis* E. J. Hill)

Jack oak, a small tree of the Lake states, is so similar in appearance to pin oak that for many years it was not distinguished from that species. Then it was discovered that the acorns of Jack oak were quite different (ellipsoidal with deep cups) and that the tree was found on dry sandy soils in contrast to the bottom-land sites of true pin oak. Jack oak has a relatively small range, extending along the Great Lakes from northwestern Ohio and Indiana through Michigan, Illinois, Wisconsin, and Minnesota to Iowa (also in Manitoba).

Blackjack Oak

(*Quercus marilandica* Muench.)

Appearance.—A small, usually scrubby tree, 20 to 40 ft. high and 6 to 12 in. in diameter, with a short trunk and narrow, rounded crown.

Leaves.—Alternate, simple, 4 to 7 in. long, 3 to 5 in. wide, broadest near the tip, tapering to a narrow, rounded or heart-shaped base, either unlobed or broadly three-lobed at the tip, shiny and dark yellow-green above, yellowish woolly below.

Flowers.—Borne separately on the same tree; the male in catkins, the female in short spikes.

Fruit.—An acorn, about ¾ in. long; the nut, enclosed in a deep bowl-shaped cup with loose reddish brown scales; kernel, bitter.

Fig. 80.—Blackjack oak. *1.* Twig × 1¼. *2.* Acorns × 1. *3.* Leaf × ½. *4.* Bark.

Twigs.—Reddish brown, stout, with star-shaped pith; terminal bud, present, narrow-conical, more or less angled, accompanied by two or three clustered laterals.

Bark.—Dark brown to black, thick, rough (blocky).

Habitat.—A typical sand-barren tree, found in open mixtures with other dry-soil species.

Distribution.—From southern New York (Long Island) to southwestern Nebraska, south to eastern Texas and central Florida.

Remarks.—Relatively rare in the north, common in the south. Of little value except for firewood.

Scrub Oak   Bear Oak
(*Quercus ilicifolia* Wang.)

A large much-branched shrub or small scraggly tree up to 20 ft. in height and 6 in. in diameter, found very commonly on the poorest of dry soils, especially in mixture with gray birch and (or) pitch pine. The leaves are 2 to 5 in. long, broadest near the tip or oblong; unlobed, three-lobed at the tip, or with 5 or 7 lobes, bristle tipped; dark green and shiny above, silvery-hairy below. The acorn is about ¾ in. long, with a deep, thick cup and usually striped nut. Important only for firewood or as a cover tree to prevent soil erosion and protect small game. Seton says it was called *bear oak* because this animal was about the only one that would eat its intensely bitter acorns.

Fig. 81.—Leaf of scrub oak × ½.

Shingle Oak   Northern Laurel Oak
(*Quercus imbricaria* Michx.)

Appearance.—A medium-sized tree, 50 to 60 ft. high and 2 to 3 ft. in diameter, with a short trunk and rounded crown.

Leaves.—Alternate, simple, 4 to 6 in. long, 1 to 2 in. wide; oblong-elliptical in outline, with smooth or slightly wavy margins and with a short bristle at the tip; shiny and dark green above, pale or brownish and hairy below.

Flowers.—Borne separately on the same tree; the male in catkins, the female in short spikes.

Fruit.—An acorn, about ¾ in. long; the nut, one-third enclosed in a thin, bowl-shaped cup; kernel, bitter.

Twigs.—Slender, greenish brown, with star-shaped pith; terminal bud, present, accompanied by two or three clustered laterals.

Fig. 82.—Shingle oak. *1.* Twig × 1¼. *2.* Leaf × ½. *3.* Acorn × 1.

Bark.—Grayish brown, with low, broad ridges and shallow furrows.

Habitat.—Found on moist hillsides or along streams in mixture with other hardwoods.

Distribution.—New Jersey to southeastern Nebraska, south to Arkansas and Georgia (below Delaware, not on the coastal plain).

Remarks.—Shingle oak belongs to a small group of red oaks that have laurel- or willowlike leaves. The only indication that the leaves might be those of an

oak is the bristle tip at the apex. Shingle oak
apparently received its name because the explorer-
botanist Michaux found the pioneers in the Ohio
Valley making split shingles, or shakes, from it.

Willow Oak

(*Quercus phellos* L.)

This oak, although found in one or more of the
states covered by this handbook, is essentially a

Fig. 83.—Willow oak. *1.* Twig × 1¼. *2.* Leaf × ½. *3.* Acorns
× 1. *4.* Bark.

southern tree, ranging along the coastal plain from
extreme southern New York to Florida, west to
Texas and north in the Mississippi Valley to southern
Illinois. Willow oak is a handsome tree as widely
used for street planting in the south as American
elm is in the north. The leaves are "willowlike,"

but more glossy, entire on the margin, with a fine bristle at the tip; the acorns resemble those of pin oak.

THE ELM FAMILY

THE ELMS

Leaves.—Alternate, simple, elliptical to oval, with coarse double teeth; more or less lopsided at the base; side veins parallel to one another.

Fig. 84.—Fruits of common elms (enlarged). Left to right, slippery elm, cork elm, American elm.

Flowers.—Perfect, borne abundantly in early spring, in clusters before the leaves.

Fruit.—Shed in late spring,[1] small, flattened, the seed surrounded by a papery margin or wing. These fruits soon die unless they alight upon moist soil where sprouting takes place within a day or two.

Twigs.—Slender, somewhat zigzag; buds, alternate; the terminal, lacking.

Remarks.—Elm wood is noted for its toughness and difficulty in splitting. Because of this, it is

[1] This is true of the three common commercial elms of the east. Two of the southern elms fruit in the fall.

generally not used as campfire material unless nothing
more tractable can be found. The bark of the elms
has been used by primitive peoples in different parts
of the world for making ropes and coarse mats (see
basswood for methods). The buds are eaten by
grouse and bobwhite; the twigs browsed by deer and
rabbits; and the fruits of some species are sought by
squirrels.

KEY TO THE ELMS

1. Leaves, more or less equal at the base (symmetrical),
   smooth or only slightly rough; twigs, usually prominently corky . . . .**Cork elm** (p. 174)
1. Leaves, lopsided at the base, moderately or very rough
   (like sandpaper); twigs, never corky **2**

2. Leaves, oval, tend to look troughlike by a fold at the
   midrib, *very* rough; outer bark, dull reddish brown
   throughout; inner bark, slippery when chewed . . . . .
   Slippery elm (p. 172)
2. Leaves, usually elliptical or broadest near the apex,
   flat, sparingly rough; outer bark shows alternate
   streaks of pale corky tissue and brown, fibrous layers;
   inner bark, not slippery . . . .**American elm** (p. 169)

American Elm   White Elm

(*Ulmus americana* L.)

Appearance.—A large tree, 60 to 80 ft. high and
3 to 4 ft. in diameter (max. 120 by 11 ft.), with a
trunk that divides into several arching limbs terminating in pendulous branchlets. The outline is
distinctive and makes this one of the most beautiful
of native trees.

Leaves.—Alternate, simple, 4 to 6 in. long, 1 to
3 in. wide, more or less elliptical, coarsely doubly
serrate, lopsided at the base; smooth or slightly
rough above, usually hairy below.

Fig. 85.—American elm.   *1.* Twig × 1¼.   *2.* Flower cluster × ¾.   *3.* Flower × 4.   *4.* Fruit × 1.   *5.* Leaf × ½.   *6.* Bark. *7.* Section through bark showing alternating brown and white layers.

Flowers.—Perfect, borne in clusters early in the spring before the leaves appear, each flower with a long slender stem.

Fruit.—Matures in the spring as the leaves unfold; about ½ in. long; seed cavity, surrounded by a thin papery wing usually notched at the apex and hairy on the margin.

Twigs.—Slender, brownish; terminal bud, lacking; the laterals, brown, acute, but not sharp pointed.

Bark.—Gray, soon furrowed, eventually divided into interlacing fibrous-appearing ridges separated by roughly diamond-shaped fissures; but sometimes without this appearance and flaky instead. When sliced with a knife, the fibrous layers are seen to be separated by light-colored, corky patches.

Habitat.—Although commonly occurring as a swamp or lowland tree, this species is cosmopolitan and is found on a large number of different sites.

Distribution.—Southern Newfoundland through southern Canada to the east slopes of the Rocky Mountains; south to eastern Texas and central Florida.

Remarks.—American elm is one of the most widespread and well known of our native trees and in the north is the favorite shade tree for street planting. The shape of open-grown trees has been referred to as that of a wineglass, a feather duster, and even a "Colonial lady upside down." None of these terms does justice to the exceeding grace and symmetry of this tree at its best. The recent introduction of the Dutch elm disease threatens the extinction of American elm even as the chestnut blight has all but wiped out the chestnut. Efforts are being made to locate and destroy all infected American elms, and friends of the tree are urged to report any instances of individuals with wilting foliage to the U.S. Department of Agriculture. The wood of this species is extremely tough and hard to split and is used for articles where this is an important feature. The bark was used by the Indians for utensils, and rope was made from the fibrous portion (see basswood). The early settlers soaked the bark in water and pulled off long flat strips for making chair bottoms. The buds

Fig. 86.—Open-grown form of American elm.

are eaten by several birds including the pinnated grouse.

Slippery Elm   Red Elm

(*Ulmus fulva* Michx. or *Ulmus rubra* Muehl.)

Appearance.—A medium-sized tree, 60 to 70 ft. high and 1 to 2½ ft. in diameter, with a short trunk and several arching limbs terminating in more or less erect branchlets.

Leaves.—Alternate, simple, 5 to 7 in. long, 2 to 3 in. wide, oval, often creased along the midrib; coarsely doubly serrate; rough above (like sandpaper); hairy below.

Flowers.—Perfect, borne in clusters in early spring before the leaves unfold; each flower with a short stem.

Fruit.—Matures in the spring as the leaves appear; about ¾ in. long; seed cavity, surrounded by a thin smooth, papery wing.

Fig. 87.—Slippery elm. *1.* Twig ✕ 1¼. *2.* Leaf ✕ ½. *3.* Fruit ✕ 1. *4.* Bark. *5.* Section through bark showing brownish layers only (compare with that of American elm).

Twigs.—Grayish, rough, moderately stout; terminal bud, lacking; the laterals, dark brown to nearly black; flower buds often orange-tipped.

Bark.—Reddish brown, furrowed, with somewhat platy ridges not interlacing as in American elm; when sliced with a knife, reddish brown, without the light-colored, corky patches of American and cork elms.

Habitat.—Mostly on rich, well-drained limestone soils but also found elsewhere.

Distribution.—Southern Quebec to southeastern North Dakota; south to eastern Texas and western Florida.

Remarks.—Slippery elm has been noted since pioneer days on account of its inner bark, which is aromatic and chewed to quench thirst, used as a poultice when powdered or in the form of a water decoction for throat inflammations and fever. The wood is similar to that of American elm; and the fibers of the inner bark furnish lacings or cordage (for preparation see basswood). The Iroquois called this elm Oo-hoosk-ah ("it slips") and in the spring peeled the bark for making canoes which were crude compared with the northern birch bark product. The buds are eaten by several birds including the pinnated grouse.

Cork Elm   Rock Elm

(Ulmus thomasi Sarg.;
Ulmus racemosa Thom.)

Appearance.—A medium-sized tree, 60 to 70 ft. high and 1 to 2½ ft. in diameter (max. 100 by 3 ft.), with an oblong crown and a trunk that extends undivided nearly to the top of the tree.

Leaves.—Alternate, simple, 2½ to 4½ in. long, 1¼ to 2¼ in. wide, more or less elliptical or broadest above the middle, coarsely doubly serrate, more inclined to be symmetrical at the base than those of the other elms, usually smooth above, hairy below.

Flowers.—Perfect, borne in the spring before the leaves, clustered along a central stem instead of all from the same point as in the other elms.

Fruit.—Matures in the spring as the leaves unfold, about ¾ in. long, hairy, with an encircling wing; seed cavity indistinct.

Fig. 88.—Cork or Rock elm. *1.* Fruit × 1. *2.* Corky outgrowths of twig. *3.* Leaf × ½. *4.* Flower cluster × ¾. *5.* Twig × 1¼.

Twigs.—Brownish; after the first year, developing corky ridges; terminal bud, lacking; laterals, brown, sharply pointed.

Bark.—Grayish, deeply furrowed, similar to that of American elm but rougher and often darker; fibrous areas interrupted by corky patches.

Habitat.—Especially common on limestone outcrops, but found on many types of soil.

FIG. 89.—Bark of cork elm.

Distribution.—Southern Quebec to southern Minnesota, south to eastern Kansas and central Tennessee (rare or not reported for wide areas included in this range).

Remarks.—This is the best of the elms commercially because of its better form and exceedingly hard, tough wood. The amount of cork on the branches is quite variable; and especially on large forest-grown trees, cork may be difficult to find. For this reason it is possible that in some places, it has not been separated from the American elm which it resembles. The fruits, the largest of the eastern elms, appear to be eaten by squirrels; in any event it is often difficult to find trees from which the seed has not been removed.

Hackberry

(*Celtis occidentalis* L.)

Appearance.—A small to medium-sized tree, 30 to 40 ft. high and 1 to 2 ft. in diameter (max. 130 by

4 ft.), with a short trunk which divides to form a bushy, more or less oblong-rounded crown.

Leaves.—Alternate, simple, 2½ to 4 in. long, 1½ to 2 in. wide, ovate to oblong-lance-shaped; sharply serrate, with a narrow curved tip and unequally heart-shaped base; smooth or slightly rough above, smooth or sparingly hairy below.

Fig. 90.—Hackberry. *1.* Twig × 1¼. *2.* Perfect flower × 3. *3.* Fruit × ¾. *4.* Pit showing network of ridges × 1. *5.* Leaf × ½. *6.* Bark.

Flowers.—Both perfect and unisexual (male and female) flowers borne on the same tree.

Fruit.—Borne on slender stems, a thin-fleshed drupe (like a cherry), about ⅓ in. in diameter, with a large pit whose surface bears a netlike pattern.

Twigs.—Reddish brown, slender, zigzag; terminal bud, lacking; the laterals, closely appressed to the twig.

Bark.—Grayish brown with characteristic **corky** warts or ridges, later somewhat scaly.

Habitat.—Commonly found on limestone outcrops, but also on many other sites.

Distribution.—New England (rare) to eastern North Dakota, south to Kansas and Virginia.

Remarks.—Although this tree is of little importance as a timber producer, its wood is cut and mixed with that of elm. The fruit is eaten by birds (as many as 25 kinds); and although good seed years may not occur often, the tree can be grown for the bird food that it furnishes. The flesh of the fruit is thin but has a pleasant datelike taste.

THE MULBERRY FAMILY

Red Mulberry

(Morus rubra L.)

Appearance.—A small to medium-sized tree, 25 to 40 ft. high and 1 to 1½ ft. in diameter (max. 80 by 7 ft.), with a short trunk and broad, round-topped crown of small branches.

Leaves.—Alternate, simple, 3 to 5 in. long, 2 to 4 in wide; oval to nearly circular, coarsely serrate; either unlobed or two-, three-, or severally lobed, (the former most common on large trees) more or less rough (like sandpaper) above, hairy below. When broken, the leaf stems exude a milky sap.

Flowers.—Male and female borne on separate trees; the former in catkins, the latter in hanging spikes.

Fruit.—Resembling a blackberry, sweet and edible; about 1 in. long; composed of many minute, closely packed drupes (like miniature cherries).

Twigs.—Slender or moderately stout, brownish; terminal bud, lacking; the laterals, ovoid; leaf scars, nearly circular or oval; bundle scars form a ring, often visible without a lens; broken twigs exude a milky sap.

Bark.—With long, somewhat flaky plates or ridges.

Habitat.—On bottom lands or rich, moist hillsides.

Distribution.—Southern New England to eastern South Dakota, south to eastern Texas and Florida.

Fig. 91.—Red mulberry. *1.* Male flowers × ½. *2.* Female flowers × ½. *3.* Fruit × ¾. *4.* Twig × 1¼. *5, 6,* and *7.* Leaves showing variation in lobing × ½.

Remarks.—Red mulberry is a relatively rare tree not often cut for lumber but one of the most valuable to plant if one wishes to attract birds and other small animals. The fruit is eagerly eaten by these visitors,

and scarcely any may be left for the nominal owner
unless he is there to prove his claim. By the same
token, a mulberry is a poor tree to plant too close to
the house unless one likes to be awakened at dawn
by noisy bird chatter. At least 21 kinds of birds,
including the bobwhite, are reported to eat mulberries which are also sought by squirrels and by
skunks.

The bark is fibrous like that of the elms, and the
southern Indians wove cloth from these fibers (see
basswood for preparation). The wood is durable
(fence posts) and was used for "tree nails" (wooden
pins) and other items in shipbuilding.

White Mulberry

(*Morus alba* L.)

The leaves of white mulberry are the chief food of
the silkworm, and the tree was introduced from
eastern Asia in colonial times when it was thought
that the silkworm industry might be established in this
country. Although the tree flourishes here and has
been widely spread by birds, the difference in the
standard of living made the raising of silkworms
not economically possible.

The leaves are usually smaller and smoother than
those of red mulberry. The fruit is paler, often white,
and the bark has a yellowish tinge especially in the
furrows.

Paper-mulberry

[*Broussonetia papyrifera* (L.) Vent.; *Papyrius papyrifera* (L.) Kuntze]

Paper mulberry comes from eastern Asia where,
for centuries, the inner fibrous bark has been used in
making paper. The leaves are unlobed, mitten-
shaped, or three-lobed and, like the true mulberries,
exude a milky juice when crushed. The twigs are

hairy. The fruit is globose, and the bark shows a network of yellowish interlacing fissures. The tapa cloth of the South Seas is made by pounding the inner bark previously soaked in water. Paper-mulberry is now found through the east as far north as southeastern New York.

Osageorange

[*Maclura pomifera* (Raf.) Schn.; *Toxylon pomiferum* Raf.]

Appearance.—A medium-sized tree (in its natural range), 50 to 60 ft. high and 2 to 3 ft. in diameter with a short trunk and irregular, rounded crown.

Fig. 92.—Osageorange. *1*. Leaf × ½. *2*. Twig × 1¼. *3*. Fruit × ½.

Leaves.—Alternate, simple, 3 to 5 in. long, 2 to 3 in. wide; ovate-lance-shaped, with an entire margin and abruptly narrowed apex; shiny and smooth above, paler and sometimes hairy below; exude a milky juice when crushed.

Flowers.—Male and female borne in globose heads on separate trees.

Fruit.—About 4 in. in diameter, a peculiar green, solid, ball-shaped multiple of small drupes, exuding a milky sap when bruised.

Twigs.—Slender or moderately stout, orange-brown; terminal bud, lacking; the laterals, small and inconspicuous, armed with sharp spines; sap, milky.

Bark.—More or less fibrous in appearance, dark orange.

Habitat.—In its natural range found in bottom lands but will grow on many soil types.

Distribution.—Originally restricted to the south-west, but now widely planted throughout the United States.

Remarks.—Because of its sharp spines this tree makes an "exclusive" hedge when planted in rows and trimmed. The bright orange wood is one of the heaviest of native timbers and was well known to the Indians as a source of bowstaves for which it is excellent (hence the French name *bois d'arc*); in the early eighteen hundreds, Bradbury said that in Arkansas, the price of a good osage bow was a horse and blanket. The chips when boiled release a yellow dye which can be used to color cloth. Unfortunately, the large fruit is not edible. Nuttall said that like paper mulberry, the bark "affords a fine white flax" (fibrous material; see basswood for preparation).

THE MAGNOLIA FAMILY

Cucumbertree

(*Magnolia acuminata* L.)

Appearance.—A medium-sized to large tree, 80 to 90 ft. high and 3 to 4 ft. in diameter (max. 100 ×

Fig. 93.—Cucumbertree. *1.* Twig × 1¼. *2.* Flower × ½. *3.* Fruiting cone × ¾. *4.* Leaf × ½. *5.* Bark.

For descriptive legend see opposite page.

4½ ft.), with a trunk extending high into the pyramidal crown which often reaches nearly to the ground.

Leaves.—Alternate, simple, 6 to 10 in. long, 4 to
6 in. wide, broadly elliptical, with an entire or slightly
wavy margin and pointed apex; smooth above, smooth
or hairy below.

Flowers.—Perfect, 2 to 3 in. long, greenish yellow,
floral parts arranged in a conelike structure enclosed
by six petals.

Fruit.—About 2½ in. long, a peculiar conelike
aggregation of units, each of which opens and releases
a bright red seed suspended on a slender thread. The
immature fruit looks something like a cucumber;
hence the name.

Twigs.—Moderately stout, reddish brown; terminal bud, about ¾ in. long, covered with silvery,
silky hairs; the laterals, smaller; leaf scars, horseshoe-shaped.

Bark.—Furrowed, with narrow flaky ridges.

Habitat.—Does best on rich, moist soils along
streams or on cool hillsides.

Distribution.—Central New York to southern
Illinois and Oklahoma, south in the Appalachians to
northern Georgia and Alabama.

Remarks.—The wood is similar to that of tuliptree
(yellowpoplar) and is used for the same purposes.
At least five species of larger birds, such as quail,
eat the seeds of this and other magnolias. Because
of its large leaves and pleasing form, cucumbertree
is planted ornamentally. According to Sargent
the exotic magnolias make best growth when grafted
on stocks of this tree.

Michaux reported that in the Allegheny region,
the early settlers collected the conelike fruits in
midsummer, steeped them in whisky, and took a glass
of this bitter liquor once a day to ward off "autumnal
fever."

Sweetbay   Swampbay

(*Magnolia virginiana* L.)

This is essentially a southern tree and reaches its
northern limit near the coast in Massachusetts and
Long Island (New York). Sweetbay is a small tree
or large shrub with narrowly elliptical or oblong
leaves and attractive creamy white, fragrant, cup-
shaped flowers 2 to 3 in. in diameter. The buds are
bright green, and the fruit is usually shorter than
that of cucumbertree. Sweetbay is a valuable
ornamental where the climate is sufficiently mild.

Tuliptree   Yellowpoplar

(*Liriodendron tulipifera* L.)

Appearance.—The tallest of eastern hardwoods,
80 to 100 ft. high and 4 to 6 ft. in diameter (max.
198 by 12 ft.), with a long columnar trunk and narrow
pyramidal or oblong crown. In the forest, the
straight, clear, massive bole, rising 60 ft. or more
before the first branch is reached, is unsurpassed in
grandeur by any other eastern broadleaved tree.

Leaves.—Alternate, simple, borne on long, slender
stems; the blade, smooth above and below, 4 to 6 in.
in diameter, circular in outline, mostly four-lobed,
broadly notched at the apex (hence the name *saddle-
leaf tree);* stipules, large and conspicuous, together
encircling the twig.

Flowers.—Perfect, about $1\frac{1}{2}$ in. in diameter, cup-
shaped; the central portion conelike, surrounded by
six greenish-yellow petals.

Fruit.—About $2\frac{1}{2}$ in. long; a conelike aggregation
of terminally winged, angled "seeds" (technically
fruits).

Twigs.—Moderately stout, reddish brown often
with a purplish bloom; pith, when sliced lengthwise,

Fig. 94.—Yellowpoplar. *1.* Twig × 1¼. *2.* Fruit × 1.
3. Fruiting cone × ¾. *4.* Flower × ¾. *5.* Leaf × ½. *6.*
Young stem showing characteristic white depressions. *7.* Bark
of old tree.

shows darker bars of tissue running crosswise; terminal bud, about ½ in. long, flattened (duck-billed in appearance); laterals, smaller; leaf scars, circular or oval.

Bark.—At first dark green and smooth, soon shows scattered whitish spots in the developing furrows, a feature that persists for many years; on old trees, heavily furrowed, grayish, especially in the bottoms of the furrows.

Habitat.—Does best on moist, deep rich soils.

Distribution.—Rhode Island to Michigan, south to Louisiana and Florida.

Remarks.—Tuliptree is one of the most important of commercial "hardwoods," although the wood is actually *soft* and easily worked. It is used for many purposes including furniture where its ability to take glue makes it an ideal veneer core upon which to lay thin sheets of more expensive woods, such as mahogany and walnut. As a firewood, it is similar to other soft hardwoods; but the heartwood is durable in contact with the soil. The name *poplar* is an unfortunate one, since this tree is not in the same family as the true poplars and cottonwoods. Besides being the tallest of eastern hardwoods, tuliptrees may also be the most massive, since reports of trees up to 16 ft. in diameter have been received; this measurement, however, has not been verified. Small animals eat the winged fruits which, in a good seed year, are cast in great numbers; four species of birds including bobwhite utilize them. Because of its pleasing appearance and almost perfect symmetry, the tuliptree is widely planted as an ornamental.

All parts of the tree are more or less bitter and slightly aromatic, especially the inner bark of the root from which hydrochlorate of tulipiferene, an alkaloid and heart stimulant, can be made.

In Virginia and Pennsylvania the tulip was called *canoe tree* because the Indians made dugouts from it.

There is but one species of tuliptree on the North American continent. The only other species in the entire world, in China, is so similar that were one to know the American form he would instantly recognize the other. Before the last ice age, one or more species were found in Europe as well as the one here. As the ice pushed slowly southward, the European trees were wiped out because of the great east and west mountain barriers and the Mediterranean Sea. But in America, the Appalachians run north and south, and trees by seeding slowly southward were able to "move" before the ice front. When the ice retreated, the trees moved northward again. This story explains why we have so many more kinds of trees in the eastern United States and Canada than are found in Europe.

THE CUSTARD-APPLE FAMILY

Pawpaw

[*Asimina triloba* (L.) Dun.]

This is a relatively rare shrub or small tree reaching a maximum height of 40 ft. and a diameter of 12 to 18 in. The leaves are oblong or broadest above the middle, 6 to 12 in. long, entire on the margin, and smooth on both surfaces. The flowers are showy, purple, about $1\frac{1}{2}$ in. in diameter, with six petals. The fruit is a large oblong berry, about 4 in. long, and contains several seeds nearly 1 in. long; the fruit is edible[1] and often appears on the local markets. Of no value as a timber species, pawpaw merits planting as an ornamental on account of its attractive flowers

---

[1] Sargent says that there are two kinds: the yellow one which is edible and the white one with a disagreeable odor. Otherwise there is no observable difference between the two forms.

and edible fruit, which is also sought by the gray fox, opossum, and raccoon.

The inner bark is fibrous and has been used for making fish nets.

THE LAUREL FAMILY

Sassafras

[*Sassafras albidum* (Natt.) Nees;
Sassafras variifolium (Salisb.) Kuntze]

Appearance.—A medium-sized tree, 40 to 50 ft. high and 1 to 2 ft. in diameter (max. 100 by 6 ft.), with a short trunk and more or less scraggly, bushy crown.

Leaves.—Aromatic, alternate, simple, 4 to 6 in. long, 2 to 4 in. wide, elliptical to oval, of three types usually on the same tree: (1) unlobed, (2) mitten-shaped, (3) three-lobed; the unlobed leaves are most common on older trees; surfaces, smooth above, smooth or sparingly hairy below.

Flowers.—Male and female borne on separate trees, in clusters, on drooping stems.

Fruit.—A dark blue drupe, about ½ in. long, borne on a red, fleshy stalk.

Twigs.—Bright green, smooth, aromatic; terminal bud, present, larger than the laterals.

Bark.—Very deeply furrowed into blocky ridges with frequent horizontal cracks ("axe marks").

Habitat.—Although doing best on rich moist soils, it is a widespread and aggressive weed tree on old fields, where, especially in the south, it may come in almost to the exclusion of other species.

Distribution.—From southern Maine to Iowa, south to eastern Texas and Florida.

Remarks.—Of little value as a timber tree, sassafras has been famous since pioneer days for the "tea" made by boiling the root bark. Oil of sassafras from the distillation of the roots and root bark is used in soaps and rubbing lotions. To chew on a twig is to identify the sassafras, since no other native tree has the identical and characteristic spicy flavor.

Fig. 95.—Sassafras. *1.* Fruit × ¾. *2.* Three-lobed leaf × ½. *3.* Unlobed lead × ½. *4.* Two-lobed or mitten-shaped leaf × ½. *5.* Twig × 1¼.

Michaux says "In Virginia and in the more southern states, the country people make a beer by boiling the young shoots of the sassafras in water, to which a certain quantity of molasses is added, and the whole is left to ferment; this beer is considered as a very salutary drink during the summer."

The Choctaw Indians of Louisiana powdered the leaves (then called "gumbo filet," "gumbo file," or "gumbo zab") and used them for flavoring and giving a ropy consistency to soup.

An extract of the bark is said to produce a good orange color on wool.

The wood of sassafras is soft and similar to that of chestnut. In like manner, it pops and shoots sparks when used as firewood. Gibson says that in Arkansas and Mississippi, bedsteads were made from sassafras to ensure sounder sleep. (Keep off evil spirits?) The wood is durable, and dugout canoes made from it have lasted 30 years or more.

The fruit is eaten by at least 18 species of birds and a few mammals.

THE WITCHHAZEL FAMILY

Sweetgum Redgum

(Liquidambar styraciflua L.)

Appearance.—A large timber tree, 80 to 120 ft. high and 3 to 4 ft. in diameter (max. 150 by 5 ft.), with a long, tapering trunk and pyramidal or oblong crown.

Leaves.—Alternate, simple; borne on long stems; the blade 4 to 7 in. in diameter, circular, deeply five- to seven-lobed ("star-shaped") with serrate margins; smooth above and below (except for tufts in the vein axils); somewhat fragrant when crushed.

Flowers.—Male and female borne separately on the same tree, in globose heads.

Fruit.—Borne on a slender stem; a woody globose head, about 1 in. in diameter, composed of many beaked capsular units, each releasing one or two

Fig. 96.—Sweetgum. *1.* Twig × 1¼. *2.* Flowers × ½. *3.* Fruiting head × 1. *4.* Seed × 1. *5.* Leaf × ½. *6.* Bark. (*Photograph by U. S. Forest Service.*)

terminally winged seeds, together with large numbers of infertile seeds resembling sawdust.

Twigs.—Greenish, moderately stout, with star-shaped pith; terminal bud, present, larger than the laterals; leaf scars, crescent-shaped to triangular; bundle scars, three, appearing as white rings, often visible without a lens; twigs, more or less corky, especially after the first year.

Bark.—Furrowed into narrow, sometimes flaky ridges.

Habitat.—A common bottom-land tree of very widespread occurrence in the south and found on many types of soil. When planted, it will grow on dry soils.

Distribution.—From southwestern Connecticut to Missouri, south to eastern Texas and Florida.

Fig. 97.—Sweetgum twig showing corky wings × ½.

Remarks.— From bark fissures or incisions, small amounts of a balsamic resin which solidifies upon drying is obtained. This is practically identical with the Oriental storax used in perfumery. During war our native tree is used.

Twelve birds including the bobwhite are reported to eat the seeds. Now that veneered furniture has so largely replaced solid pieces, redgum wood has come into prominence. It can be stained in imitation of more expensive woods, and frequently trees are found that produce a veneer pattern of great beauty. In the north, the tree is often planted as an ornamental; and when the autumn frosts come, the leaves turn to a bright scarlet.

Witchhazel

(*Hamamelis virginiana* L.)

For the most part, a shrub, this species sometimes becomes a small tree, especially in the south. It is one of the most unusual of our native woody plants, with its wavy-margined, lopsided leaves, flowers with long narrow yellow petals appearing in the autumn, half-moon-shaped tawny buds, and woody capsular fruit which opens slowly and then forcibly

Fig. 98.—Witchhazel. *1.* Flowers × 1. *2.* Fruit and seeds × ¾. *3.* Bud × 2. *4.* Leaf × ½.

ejects the shiny black seeds some distance from the parent shrub. If one brings in a fruiting branch in autumn, he will be startled from time to time to hear a loud snap accompanied by the flying seed being catapulted through the air. Observation seems to show that this is done in the same way that a pumpkin or watermelon seed can be released from the thumb and forefinger (familiar to most of us from childhood).

Witchhazel is one of the species selected by "water diviners" from which to cut their forked sticks. Surprising as it may seem, there are many people who still believe, in this twentieth century, that water can be found by walking along with the forked stick properly balanced between thumb and fingers and waiting until it suddenly dips forward. The author hopes to find someone who will come forward, allow himself to be blindfolded, and be guided here and there to locate a "vein" of water. If the stick bobs down every time he is led over a certain place, it will at least call for an explanation. Meanwhile there seems to be no possible connection between the antics of a witchhazel or any other kind of fork and the presence of subterranean springs. Yet how can it be *disproved* when the well builder digs where he is told and water springs forth? Sargent took no stock in such magical properties, since he says, "the popular name of this plant is due to the fact that it was early used by impostors to indicate the presence of precious metals in the soil and to discover springs of water. For this purpose, a forked branch is twirled between the fingers and thumbs of the two hands; then at the place where the fork points, water or gold is declared to exist."

The Indians are said to have made from the inner bark a poultice for inflamed eyes and skin surfaces, a very interesting use considering that we now use an alcoholic extract of this tree for similar purposes.

Seton says that a snuff prepared by powdering the dried leaves is good to stop bleeding from small cuts.

THE PLANETREE FAMILY

Sycamore   Planetree

(*Platanus occidentalis* L.)

Appearance.—A large timber species (probably with the greatest diameter of any eastern tree), 80 to 100 ft. high and 3 to 8 ft. in diameter (max. 175 by 14 ft.), with a continuous tapering trunk or

For descriptive legend see opposite page.

breaking up near the ground into several large limbs terminating in crooked branchlets. The crown is open and spreading.

Leaves.—Alternate, simple, borne on slender stems; the blade, 4 to 7 in. in diameter, three- to five-lobed, with the margins coarsely toothed; smooth above, hairy along the veins below; stipules, leaflike and conspicuous, together extending completely around the twig; base of leaf stem, hollow, enclosing next year's bud (the only eastern species with simple leaves having this feature).

Flowers.—Male and female borne separately in globose heads on the same tree.

Fruit.—A globose brownish head about 1 in. in diameter borne on a long woody stem; the small unit fruits ("seeds"), each with a circle of hair at the base, become detached at maturity or during the winter and following spring.

Twigs.—Brownish, slender, zigzag; terminal bud, lacking; the laterals, conical, each with a single outer scale, sticky within; leaf scars, surrounding the buds; stipule scars, encircling the twig.

Bark.—The most striking feature of the tree; mottled creamy white and brown, the former where the old bark peels off; at the base of old trees, almost entirely brown and flaky.

Habitat.—One of the commonest of stream-bank and bottom-land trees scattered in mixture with other hardwoods.

Distribution.—Southern Maine to Iowa, south to Texas and northern Florida.

Remarks.—In appearance, this is perhaps the most distinctive of eastern hardwoods with its massive

---

Fig. 99.—Sycamore. *1*. Leaf × ½. *2*. Head of female flowers × ½. *3*. Head of male flowers × ½. *4*. Fruiting head × ¾. *5*. Fruit × 1. *6*. Twig × 1¼. *7*. Bark of young tree. *8*. Bark of old tree.

mottled trunk and "button balls" swinging high in the air.

Both Sargent and Michaux mention that some people believed that the down from the young leaves if breathed would produce irritation of the lungs; but the latter says "the slightest zephyr suffices to waft to a distance and to disperse in the airy waste this light and impalpable substance." He also wrote that the Illinois French made dugout canoes of sycamore, one of which, 65 feet long, carried 9,000 lb.

The small fruits are eaten by a few birds; deer sometimes browse the twigs; and muskrats are known to chew the bark. The wood is tough, very difficult to split, and has been used for such items as butcher's blocks where this feature is of importance.

London Plane

(*Platanus acerifolia* Willd.)

A hybrid, presumably between the native sycamore and the Oriental plane, this tree is a common ornamental. The freshly exposed bark is olive-green. The leaves are smaller and smoother than in sycamore, and usually two of the globose heads are borne on a single stem.

THE ROSE FAMILY

THE CHERRIES[1]

Leaves.—Simple, alternate, lance-shaped to elliptical or widest above the middle, serrate.

Flowers.—Perfect; borne either in clusters on stems of about the same length (common sweet cherry) or on stems attached alternately to a central one (black cherry).

Fruit.—A drupe (a cherrylike fruit).

[1] Included in the same group botanically are the peach and plum.

Twigs.—Moderately slender, with a typical bitter-almond flavor; buds, alternate; the terminal, present.

Bark.—At first smooth, usually with conspicuous lenticels; later, in some species, becomes scaly.

Remarks.—As a group, the cherries furnish important food for wild life. The fruit is eaten by 70 or more kinds of birds; bears break down the bushes to feed on cherries; and skunk, raccoon, and red fox also seek them. The branchlets are browsed by deer, and rabbits chew the bark of some species. The flowers furnish pollen and some nectar to honeybees. Wilted foliage and twigs are poisonous for cows to eat.

KEY TO THE CHERRIES

1. Leaves, elliptical to lance-shaped.................... **2**
1. Leaves, broader, usually oval or broadest above the middle.. **3**

    2. Leaves, leathery, dark green, usually with hair along the midrib near the base, turning reddish brown as the leaf matures; fruit, nearly black, borne on stalks along a central stem.............
                  Black cherry (p. 199)
    2. Leaves, thin, yellow-green, not hairy below; fruit, red, borne in clusters on stalks of about the same length......................**Fire cherry** (p. 203)

3. Teeth, sharp; a native shrub or small tree; fruit, borne on stalks along a central stem, puckery; stem, exceedingly bitter.................. **Choke cherry** (p. 202)
3. Teeth, often rounded; introduced tree now growing wild; fruit, clustered on stalks of about the same length, edible; large lens-shaped lenticels on the bark........
        Common sweet, or **Sour cherry** (p. 204)

Black Cherry

(Prunus serotina Ehrh.)

Appearance.—A medium-sized timber tree, 50 to 60 ft. high and 2 to 3 ft. in diameter (max. 100 by

Fig. 100.—Black cherry. *1.* Twig × 1¼. *2.* Flower cluster × ½. *3.* Fruits × ½. *4.* Leaf × ½. *5.* Bark of young tree. *6.* Bark of old tree.

5 ft.) often with a short trunk and irregular, rounded crown.

Leaves.—Alternate, simple, 2 to 6 in. long, 1 to 1½ in. wide; oblong-elliptical to nearly lance-shaped, finely serrate with callous, incurved teeth; shiny and dark green above, paler below and usually densely hairy along the midrib at the base of the leaf (pale hair turning reddish brown at maturity).

Flowers.—Perfect; about ¼ in. wide, arranged on short stems along a central axis.

Fruit.—A dark red, nearly black drupe about ⅓ in. in diameter; edible.

Twigs.—Slender, reddish brown, often with a peeling grayish skin; taste like bitter almonds when chewed; terminal bud, present, larger than the laterals.

Bark.—At first smooth, red-brown to nearly black, marked with long conspicuous linelike horizontal lenticels; later breaking up into scaly plates, eventually entirely scaly.

Habitat.—Very widely spread, although does best on moist, deep, rich soils.

Distribution.—From Nova Scotia to North Dakota, south to Texas and Florida.

Remarks.—Black cherry is one of our finest and most valuable forest trees. The hard wood is of high quality for use in furniture, interior finish, and similar purposes and as firewood produces a bed of hot coals. An extract from the bark is used in medicine as a sedative or tonic; and when the tree is covered with bloom, it has a high ornamental value. The fruit is sometimes used to flavor rum or brandy, which then becomes "cherry bounce." It goes without saying that the cherries are eagerly sought by the birds; in fact, so much so that it was practically impossible to collect a complete cluster for photographing. Squirrels eat the seeds, and rabbits gnaw the bark in winter.

Chokecherry

(*Prunus virginiana* L.)

Although commonly a shrub, this species may attain tree size and very rarely reaches a height of 35 ft. and a diameter of 12 in. The leaves are usually broadest above the middle and sharply serrate on the margin (poisonous to eat when wilted). The flowers and fruit are similar in appearance to those of black cherry, but the fruit of chokecherry, as its name suggests, is very puckery except when wholly ripe.[1] Then it has its devotees, even though the astringency is still not entirely lacking. The twigs have even a ranker odor and taste than those of the other cherries, and the bud scales in contrast are gray margined. The bark does not show the conspicuous lenticels of the black cherry.

Fig. 101.—Choke-cherry. *1.* Leaf × ½. *2.* Bud × 1.

It would be hard to find a more widespread and hardy shrub or small tree than chokecherry. With its varieties, it covers nearly all of tree-habitable North America exclusive of Mexico, a few of the western

[1] Wood in *New England's Prospect* says "The cherrie trees yeeld great store of cherries which grow on clusters like grapes; they be much smaller than our English Cherrie, nothing neare so good; if they be not very ripe, they so furre the mouth that the tongue will cleave to the roofe, and the throate wax horse with swallowing those red Bullies (as I may call them) being little better in taste. English ordering may bring them to be an English cherrie, but yet they are as wilde as the Indians."

states, and Florida. It is found almost anywhere, especially along fence rows or where birds may be in the habit of roosting, and when once established defies extremes of climate and the ravages of rabbits who have little taste for its bitter twigs[1] although ready to eat the needles and even twigs of planted conifers (pines) near by. In frost pockets in the Adirondacks, where the temperature sinks to 40° below zero or lower, chokecherry thrives to the exclusion of more useful woody plants. It is indeed unfortunate that this species is at present only a weed tree, but perhaps with developing interest in soil-erosion control and game management it can be utilized. It appears to have many of the necessary qualifications.

Fire Cherry   Bird Cherry   Pin Cherry
(*Prunus pennsylvanica* L.)

As suggested by one of its names, this small tree is common after fires and occurs in scattered mixtures with the aspens and gray birch. The leaves are narrow like those of black cherry, but lighter green, thinner, not reddish hairy below, and more inclined to be lance-shaped. The flowers are borne on stems attached at the same point instead of being arranged along a central stem, and the sour but edible fruit is bright red in color. The bark is smooth, dark red to nearly black, and marked by conspicuous, wide, lens-shaped lenticels. The tree is distinctly a forest weed, since it deteriorates and dies when from 4 to 6 in. in diameter, although sometimes reaching a greater size. Its chief function from the forester's standpoint is to prevent erosion and furnish a light

[1] At least one reliable authority says that rabbits chew the bark, but the author thinks that they do so as a last resort if nothing else is available.

Fɪɢ. 102.—Fire cherry flowers and leaves × ¾.

shade for other trees growing underneath.   Of course,

the fruit is eaten by birds (23 kinds reported) which in this way distribute the pits; and beaver eat the bark when aspen is scarce. Fire cherry ranges through Canada, the Lake states, the northeast, and the Appalachian Mountains to Georgia.

Sweet Cherry

(Prunus avium L.)

Introduced from Europe, this is now a common escape, widely distributed by birds. The leaves are about 4 in. long, broadly oblong or broadest above the

Fɪɢ. 103.—Sweet cherry. Leaf × ½. Buds × 1.

middle, with coarse, often somewhat rounded teeth

and a rounded or heart-shaped base. The flower stalks radiate from a common point like those of fire cherry, and the fruit is the familiar sweet cherry of our orchards, the quality of which, however, must be maintained by grafting. The bark is smooth, dark reddish brown, and marked by conspicuous lens-shaped lenticels.

Sour Cherry

(*Prunus cerasus* L.)

This tree is similar in general appearance to sweet cherry, but its leaves are inclined to be more elliptical in shape with narrower bases. It is another naturalized tree from Europe and in one or more of its varieties produces the common canning cherry of our markets.

Fig. 104.—Leaf of peach × ½.

PEACH AND PLUM

Peach

[*Prunus persica* (L.) Batsch.]

Supposedly from eastern Asia, this tree has been cultivated since colonial times for its fruit and is now found occasionally as an escape. The leaves are alternate, about 4 in. long, and narrowly elliptical to lance-shaped and have a tendency to be folded along the midrib and considerably wrinkled so that it is difficult to make them lie flat.

Garden Plum

(*Prunus domestica* L.)

Introduced as a fruit tree from the Old World, plum is now found growing wild in many sections of

North America. The leaves are alternate, about 2½ in. long, and ovate to oval or broadest above the middle; the veins are more or less sunken above, and the undersurface is soft-hairy.

FIG. 105.—Leaf of common plum × ½.

Native Plums

There are several native species in this group, all relatively rare large shrubs or small trees. For these, such manuals as *Trees of Northeastern United States* by H. P. Brown and *Trees of the Northern States and Canada* by R. B. Hough may be consulted.

Thornapple   Hawthorn

(*Crataegus* spp.)

Some 800 species of this group are said to be native to North America;[1] furthermore, considerable evolution is taking place so that the number of different forms is increasing. For this reason, only the general features are given, and anyone interested in going further may secure such volumes as *Manual of Trees of North America* by C. S. Sargent.

The leaves are alternate, simple, and variable in shape and character of margin (mostly irregularly toothed); the flowers are perfect, borne in showy clusters; and the fruit is a small reddish pome (like an apple). The twigs are mostly armed with formidable thorns, and the small rounded buds are covered by thick, often reddish, fleshy scales.

The thornapples are ubiquitous pasture "weeds," and a constant battle is necessary to keep them from

[1] Other estimates vary from 100 to 1,200 species!

preempting grazing lots. As seen in the accompanying illustration, cattle (also deer) browse the young growth and in this way produce very bizarre forms. The shape at first is pyramidal (see background), but the central branches eventually grow beyond the reach of the animals and bush out overhead. The fruits are eaten by at least 36 kinds of

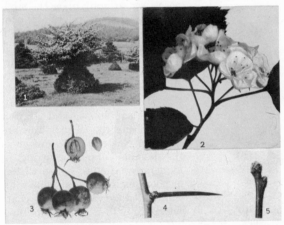

Fig. 106.—Thornapple. *1.* Pasture trees showing effects of grazing; note small pyramidal one in background. *2.* Flowers × ¾. *3.* Fruits and seed × ¾. *4.* Thorn × 1. *5.* Twig × 1½.

birds, the gray fox, and probably other animals. Because of their abundant flowers and bright red fruit, various species are planted as ornamentals, especially the English hawthorn which has deeply lobed leaves.

Wild Apple

(*Malus pumila* Mill.)

Like most of our other fruit trees, the apple was brought over in pioneer days from the Old World.

One of the most colorful frontier characters was "Johnny Appleseed," and many of the orchards through the Ohio Valley are supposed to have been started from seed left by the wandering Johnny as he stopped here and there, planting as a token the seeds that he had collected "back East" and always carried with him in a pouch. Of course, it was necessary to graft upon the seedling stock, since relatively poor fruit is produced by trees grown from seed.

The leaves are about 3 in. long, quite variable in shape but mostly elliptical to oval or broadest above the middle, with long, stout, velvety leaf stems, and grayish velvety beneath. The flowers are borne in clusters, and the fruit is the familiar apple occurring in numerous varieties, but small and of poor quality on trees gone wild. The different varieties are characterized by their fruit rather than by differences in leaves, twigs, bark, or other features. Apple twigs and buds are more or less velvety; and when chewed, the former has a characteristic "sweetish" taste.

Fig. 107.—Common apple. *1.* Leaf × ½. *2.* Bud × 1.

Flowering Crab   American Crab

[Malus coronaria (L.) Mill.]

Shrubby or of small tree stature, this species is relatively rare, occurring from New York to Wisconsin south to Missouri and the Carolinas. The leaves are alternate, ovate, and sharply irregularly serrate; the exceedingly fragrant flowers are borne in clusters;

and the sour fruit is about 1¼ in. in diameter. Because of its flowers and later fruit which may be used for preserves, the tree is sometimes planted ornamentally.

Locally where it occurs in any quantity, it furnishes important game food. Birds (38 kinds), skunks, and other animals eat the fruit, and some of the birds also eat the buds. Two or three other species of wild apple occur in the Mississippi Valley and the southern states. (See *Trees of the Northern States and Canada* by R. B. Hough.)

Fig. 108.—Leaf of flowering crabapple × ½.

Wild Pear

(*Pyrus communis* L.)

The pear was introduced from the Old World and occurs here as an escape. The leaves are alternate, about 3 in. long, oval to ovate, thin and firm, smooth on both surfaces, and borne on long, slender, smooth leaf stems. The flowers are borne in clusters, and the fruit is the pear of our orchards, although when growing wild it tends to lose its typical pear shape and becomes irregularly globose. The twigs are smooth, with conical terminal buds and sharp thorns. The twigs have a taste similar to those of the apple tree.

Fig. 109.—Leaf of common pear × ½.

Fɪɢ. 110.—American mountainash. *1*. Leaf × ½. *2*. Fruit cluster × 1. *3*. Bud × 1½.

American Mountain-ash

(*Sorbus americana* Marsh.)

A large shrub or small tree, this common North
Woods species ranges from Newfoundland to Manitoba, south through the upper Lake states and along
the Appalachians to Tennessee. The leaves are
alternate, compound, about 10 in. long and consist
of 13 to 17 oblong lance-shaped leaflets. The small
white flowers, borne in dense flat-topped clusters,
are succeeded by the bright red fruits about ¼ in.
in diameter (technically a pome, like apple). Because
of its attractive flowers, fruit, and leaves, it is much
transplanted and used as an ornamental. The name
ash as applied to this species is unfortunate, since it
is in the rose rather than the ash family.

The fruit is eaten by about a dozen species of birds,
and deer and moose browse the twigs. It is claimed
that when this and the European mountain-ash are
grown together, the birds eat the fruit of the native
tree first.

Northern Mountain-ash

(*Sorbus decora* Schneid.)

Similar in general appearance to the preceding
species, this form has larger, more showy flower
clusters; wider, more oblong leaflets; and larger
fruit (⅓ in. in diameter). It ranges from Labrador
to Minnesota, south through northern New England
and northern New York.

European Mountain-ash   Rowan-tree

(*Sorbus aucuparia* L.)

The rowan-tree, introduced from Europe, is the
commonly planted ornamental of the mountain-ash
group and is now found growing as an escape in the

north. The leaves are alternate, with 9 to 15 oblong-elliptical leaflets paired along a more or less hairy principal stem. The flowers and fruit are similar to those of the preceding species, but more showy.

According to Nuttall the rowan was planted as a "sacred" tree in the churchyards of Wales; and on a certain day each year, everyone wore a cross made from the wood to ward off evil spirits. With this as a background, the following statements seemed plausible. George Emerson says that in *Macbeth* "the sailor's wife, on the witches requesting some chestnuts, hastily answers 'A roan-tree (rowan) witch'! but all the editions have 'Aroint thee witch'! which is nonsense and evidently a corruption." For this ingenious idea, most authorities have nothing to say. In Cassell's *Illustrated Shakespeare*, "Aroint" is traced to the Latin *averrunco* meaning "I drive away evil" or "stand off," "begone." It is stated that in Cheshire, England, if a cow crowds the milkmaid too closely, she will give it a push, at the same time exclaiming "Roint thee" (stand off). It is not easy to see how in such an instance, particularly, "roan-tree" could be substituted and still make sense. This

Fig. 111. — Rowan - tree or European mountain - ash. *1.* Leaf × ½. *2.* Bud × 1½.

story is a clear illustration of how fables regarding trees and other things often get started.

Shadbush   Serviceberry

[*Amelanchier arborea* (Michx. f.) Fern.]

Usually a large shrub or small tree, shadbush may attain a maximum height of 50 ft. and a diameter of 2 ft.   The leaves are about 2½ in. long, widely elliptical or broadest above the middle; sharply serrate, rounded or heart-shaped at the base; smooth above and also below except along the veins.   The flowers o c c u r in conspicuous masses before or as the leaves unfold and at the time in the spring (April and May) when the shad ascend the New England rivers to spawn; hence the common name.   If any of the fruit is left by the birds (40 species eat it) and other animals including the skunk, red fox, raccoon, and bear, it matures in late June (also known

FIG. 112.—Shadbush.   *1.* Leaf × ½.   *2.* Bud × 1¼.

as Juneberry) and is an edible berrylike purplish pome. The Cree Indians used the fruit fresh or dried and made with it a pudding very little inferior to plum pudding, according to Emerson.   The bark except on old trunks is smooth and bluish gray.   One of the best features for recognizing the tree is the terminal bud, which is long and tapered like that of beech but with fewer scales.   When chewed, the twigs have a faint bitter-almond taste.   Because of its attractive flowers, shadbush is sometimes used ornamentally.   The wood is hard and heavy, but little used on account of the small size of the tree.

Shadbush ranges from Maine to Iowa, south to Louisiana and Missouri.

Smooth Northern Shadbush

(*Amelanchier laevis* Wieg.)

This species is very similar to the preceding but has leaves that are smooth from the beginning (those of common shadbush are silvery woolly when first unfolded), larger more showy flowers, and better tasting fruit. It is found from Newfoundland to Michigan, south to Kansas and Georgia.

THE BEAN OR PEA FAMILY

Honeylocust

(*Gleditsia triacanthos* L.)

Appearance.—A medium-sized tree, 70 to 80 ft. high and 2 to 3 ft. in diameter (max. 140 by 6 ft.), with a short trunk, often clothed with formidable branched thorns, and an irregular, open crown.

Leaves.—Alternate, compound, and also twice compound on the same tree, 6 to 10 in. long; smooth above and below; leaflets, 1 to 2 in. long, elliptical, acute or rounded at the tip, and finely toothed on the margin.

Flowers.—Small, greenish; both male flowers and perfect flowers found on the same tree; borne in separate clusters on stalks arranged along a central stem.

Fruit.—A strap-shaped, curved, twisted, mahogany-red pod, about 12 in. long, containing a number of oval seeds.

Twigs.—Zigzag, shiny, reddish brown, with three-branched (less commonly single or once-branched)

Fig. 113.—Honeylocust. *1.* Twig × 1¼. *2.* Thorn × 1¼. *3.* Bicompound leaf × ½. *4.* Once-compound leaf × ½. *5.* Bark.

For descriptive legend see opposite page.

thorns; terminal bud, lacking; laterals, partially buried in the twig tissues.

Fig. 114.—Fruit and seed of honeylocust × ⅔.

Bark.—Dark brown to nearly black, broken up into long, flat, sometimes scaly ridges, and bearing few to numerous thorns.

Habitat.—A typical bottom-land tree but thrives when planted on drier soils.

Distribution.—From Pennsylvania to Nebraska, south to eastern Texas and Alabama. Now widely planted throughout the eastern states and Canada.

Remarks.—Although inconspicuous, the flowers furnish nectar for honeybees. The green pods contain a sticky substance between the seeds which tastes like a mixture of honey and castor oil; however, cattle relish these fruits, showing that one cannot be guided by human taste in determining what animals will eat. The dry pods often remain on the tree during part of the winter and, when the wind blows, give off a dismal crackling sound as they strike one another. Some are blown off and roll considerable distances over the snow, in this way carrying the seeds to places where they may grow. According to VanDersall, root nodules (containing nitrifying bacteria), so common and important to soil improvement and typical of plants in this family, are lacking in honeylocust. The seeds (beans) are emergency food for the bobwhite and are also eaten by the gray squirrel. The wood is hard, heavy, and durable and used for fence posts and railroad ties.

Redbud   Judas-tree

(*Cercis canadensis* L.)

The redbud is one of our most beautiful native ornamental small trees, especially when planted among conifers where its masses of pink bloom contrast with the dark green background. The pealike flowers appear before the leaves or just as they are unfolding and in such quantities that the result is spectacular. The leaves are alternate, nearly circular, heart-shaped at the base, and entire on the margin; the leaf stem has a bulbous swelling at each end. The fruit is a thin, flat pod with sloping ends.

According to an ancient myth, Judas hanged himself on a tree of the Asiatic species of this group. The bloom is supposed to have been white before this took place; but in shame the tree blushed, and the

Fig. 115.—Redbud. *1.* Leaf $\times \frac{1}{2}$. *2.* Twig $\times 2$. *3.* Fruit $\times \frac{2}{3}$.

flowers have been red ever since.[1] Three kinds of birds including the bobwhite eat the seeds.

Redbud is found from New Jersey to Iowa, south to Texas and Florida; although usually of small stature, it may reach a height of 50 ft. and a diameter of 18 in.

[1] There is "still" a white-flowered form, of both the Asiatic and native trees.

Kentucky Coffeetree

[Gymnocladus dioicus (L.) K. Koch.]

Appearance.—A medium-sized tree, 60 to 80 ft. high and 1 to 2 ft. in diameter (max. 110 by 4 ft.), with a short trunk and several large limbs terminating in stout, contorted branchlets.

Leaves.—Alternate, twice compound, 1 to 3 ft. long, with upward of 100 or more leaflets arranged on

Fig. 116.—Kentucky coffeetree. *1.* Twig × 1¼. *2.* Fruit and seed × ½.

the branched midrib (rachis); leaflets, ovate, about 2 in. long, entire, and smooth on both surfaces.

Flowers.—Male and female flowers borne separately in drooping, branched clusters on separate trees.

Fruit.—A woody, thick, wide pod, about 6 in. long (less in the north) containing several hard-shelled, dark-brown seeds.

Twigs.—Stout, crooked, with large salmon-colored pith; terminal bud, lacking; laterals, minute, sunken in the twig, and surrounded by a hairy crater; leaf scars, roughly heart-shaped, large, and conspicuous.

Bark.—Peculiarly and characteristically scaly, reminiscent of that on old black cherry and yet different.

Habitat.—Mostly on low rich bottom lands, but when planted thriving on poorer drier soils.

Distribution.—From western (possibly central) New York to southern Minnesota, south to Oklahoma and Tennessee.

Remarks.—Relatively rare in the natural state, Kentucky coffeetree is much planted as an ornamental. In autumn, the leaflets fall first, leaving the extensive framework of the twice-compound leaves on the tree. At this stage, some bizarre patterns are thus formed, some appearing like giant corded spider webs. Soon the framework falls, leaving the stout, naked-appearing branches (*Gymnocladus* means "naked branch") which do not leaf out in the spring until most other trees have already done so. That such a huge leaf can develop so quickly from such an insignificant bud is nothing short of amazing. Some say that the early pioneers used the beans as a substitute for coffee, but others claim that they are very bitter and that the name was applied because of the similarity in *appearance* between the coffee bean and the seed of this tree. Probably the latter view is the correct one, although there is no doubt that the coffeetree beans were *tried* in some cases (Michaux).

Black Locust[1]

(*Robinia pseudoacacia* L.)

Appearance.—A medium-sized tree, 40 to 60 ft. high and 1 to 2 ft. in diameter (max. 100 by 3 ft.)

---

[1] Clammy locust (*R. viscosa* Vent.) is a southern Appalachian tree somewhat like black locust, except for weaker spines, sticky twigs, roseate flowers, and often hairy leaves (undersurface). It is widely planted as an ornamental.

Fig. 117.—Black locust. *1.* Twig ✕ 1¼. *2.* Leaf ✕ ½. *3.* Flower cluster ✕ ½. *4.* Fruit ✕ ¾. *5.* Fruit with one valve removed showing seeds ✕ ¾. *6.* Bark.

with a short trunk dividing into several stout branches which form a narrow or irregular open crown.

Leaves.—Alternate, compound; about 10 in. long and composed of 7 to 19 leaflets which are 1½ to 2 in. long, elliptical, entire on the margin, often notched at the apex, with a whiskerlike tip; the surface is smooth and dark green above, paler beneath.

Flowers.—Perfect, borne on stalks along a central stem; about 1 in. long; white and fragrant.

Fruit.—A brown, flat pod about 3 in. long, enclosing several kidney-shaped seeds.

Twigs.—Dull brown, zigzag, usually but not always with paired spines; terminal bud, lacking; laterals, not visible but buried in the twig tissues beneath the left scars which show cracks on their surfaces.

Bark.—Heavily furrowed into interlacing fibrous ridges.

Habitat.—Does best and suffers the least injury from the locust borer on fertile, moist soils especially of limestone origin.

Distribution.—Originally restricted to the Appalachians, this tree is now widely distributed over the eastern United States through planting, and now spontaneously.

Remarks.—Black locust produces a heavy, hard, yellowish-brown wood which shrinks or swells very little with changes in moisture content. Hence it is used almost exclusively for such items as pins to support glass insulators and "tree nails" (dowels for holding together wooden ships). It is supposed that the Indians used the wood for bows and eventually spread black locust beyond its natural Appalachian range. The tree was found planted by their houses near the Virginia coast (Sargent). They also made a blue dye from the leaves.

VanDersall says that the young shoots and bark are poisonous to all livestock, but not the leaves. Muenscher lists black locust in his book of *Poisonous Plants* and mentions that children have been poisoned by chewing the *leaves*. The seeds also are poisonous.

Several kinds of birds eat the seeds if nothing better is available. The tree is being planted extensively for soil-erosion control, since it not only holds the earth from washing away but also enriches it through the nitrifying bacteria found in granular nodules on its roots. This species was early introduced into Europe (about 1600) and is now a common tree there. There are several varieties of black locust, and especially valuable is the shipmast locust which is superior in form to the ordinary kind.

Yellowwood

[*Cladrastis lutea* (Michx.) Koch.]

As a wild tree, this is one of the rarest, found originally centered around western Tennessee and southwestern Missouri. Now, however, it is a common ornamental in much of eastern North America. The leaves are alternate, compound, with oval leaflets; the central stem hollow at the base and enclosing the next year's bud. The white fragrant flowers are borne in beautiful pendent clusters and are sought by bees and other insects. The fruit is a flat, irregularly sided pod, many of which remain on the tree until the following spring. The bark is gray like a beech; the heartwood is a bright yellow, and a dye of this color can be made from it.

THE RUE FAMILY

Northern Prickly-ash

(*Zanthoxylum americanum* Mill.)

Prickly-ash (not a true ash) is usually a medium-sized to large shrub occurring in thickets; rarely, it becomes a small tree, about 20 ft. high.   The leaves look something like those of black locust, but the

Fig. 118.——Prickly-ash. *1.* Leaf × ½.   *2.* Twig × 1½.

scattered spines on their central stems and strong lemon odor when crushed set them apart as distinctive.   Like black locust, the twigs exhibit sharp, woody, paired spines, but the red, woolly buds are strikingly different (in black locust the buds are buried in the twig and do not emerge until spring). The small, reddish fruiting capsules are aromatic with a lemon odor.

Prickly-ash is common locally in moist woods from southern Canada to Minnesota, south to eastern Kansas and Virginia.

One of the most interesting fall visitors on this shrub is the caterpillar of the giant swallowtail butterfly (*Papilio cresphontes*). It has the appearance of a piece of bird manure and, when disturbed, throws back its head and shoots out a pair of red glandular "horns" which exude an odor like putrid orange or lemon peels. Found originally in the south in orange groves, it has worked slowly north, interestingly enough on the leaves of this shrub which is in the same family as the orange and lemon.

The bark contains an alkaloid (xanthoxylin) used in medicine, sometimes for toothache, hence another common name, *toothache tree*.

Hoptree   Wafer-ash

(*Ptelea trifoliata* L.)

This is a small tree with alternate, usually trifoliate (three leaflets) leaves. The leaflets are mostly smooth margined and have a strong, somewhat orange odor when crushed. The fruit resembles that of the elms, with an encircling papery wing, but in contrast contains two seeds instead of one. The bitter fruit has been used in place of hops for brewing beer. Hoptree ranges from Pennsylvania and extreme southern New York to Michigan and Iowa, south to Texas and Georgia.

Fig. 119.—Hoptree. *1*. Leaf × ½. *2*. Fruit × 1.

THE QUASSIA FAMILY

Tree-of-Heaven   Chinese-sumac

[Ailanthus altissima (Mill.) Swing.]

The introduced tree-of-heaven, or ailanthus, is a remarkable species from the standpoint of hardiness and rapidity of growth. Once it gets a foothold, it will withstand perhaps a greater handicap in the way of poor, hard, tramped, or sunbaked earth than any native tree (it has even been found growing between

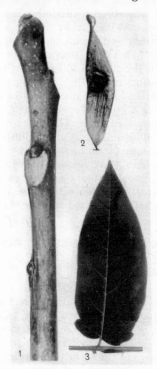

Fig. 120. ——Tree-of-heaven.   *1.* Twig   × 1¼.   *2.* Fruit   × 1.
3. Leaflet and portion of mid-rib × ½.

bricks near a second-story window in an old building). When cut, this tree sprouts with great vigor, sometimes to a height of 12 ft. the first season, and it is practically impossible to get rid of it unless the entire root system is grubbed out.

The large, compound, alternate leaves with 13 to 41 leaflets, toothed only at the base, are instantly recognized by their powerful odor when crushed or picked from the stem. This odor is something like that of popcorn with rancid butter poured over it; one of the author's students suggested that it "smelled like a zoo," not a bad comparison. Certain trees bear only male flowers (also odoriferous); others also have perfect ones which develop the fruit. The fruit often occurs in great masses on the tree and looks something like a crude propeller, with a twisted wing and central seed cavity. The twigs are stout, with large conspicuous leaf scars and reddish pith.

If you plant this tree, be sure that you really want it (which is doubtful) because it will defy you to get rid of it, once started. Tree-of-heaven may even become a forest "weed" within the next 25 years. The wood is soft, weak, not durable, and apparently of little use.

THE CASHEW OR SUMAC FAMILY

Staghorn Sumac[1]

(*Rhus typhina* L.)

So called probably because the bare branches look like a stag's horns in the velvet, this large shrub or small tree is one of the commonest woody plants ranging through southern Canada and the northern United States (Appalachians to Georgia).

The alternate compound leaves with toothed leaflets, silvery below, exude a milky juice when broken. Since male and female flowers are borne on separate trees, fruit is found only on the latter.

[1] Smooth sumac (*R. glabra*) is not so common as staghorn sumac. Smooth sumac has smooth leaves and twigs and more open fruiting clusters. Otherwise the two species are similar. The Indians ate the young sprouts raw, as a salad.

The compact, bright red, conelike fruit clusters are very conspicuous and consist of many small drupes covered with red, glandular, acid-containing hairs.

FIG. 121.—Staghorn sumac. Twig × 1¼. Fruit × ½. Leaf × ½.

The taste is something like that of a sour apple (not unpleasant) and on a hot day not bad as a thirst quencher; of course after sucking out the juice, the remainder is discarded. The fruit can also be used to make what some call *Indian lemonade*.

The bark and leaves contain tannin which is sometimes used, and Seton says that a black ink can be made by boiling down the leaves and fruit in water. It has been found that a few drops of an iron (ferric) salt greatly improve the quality and prevent the solution from molding, which it otherwise soon does. Sargent mentions that pipes (tubes) were made by punching the pith out from the stout twigs; these were used for drawing spring sap from sugar maple. One of these tubes, if made long enough, is what Stewart Edward White calls a "persuader," an instrument to blow through for "brightening up" a slow, perhaps dying campfire.   It is better than waving a hat at it or blowing "wholesale," since the stream of air is under control and can be directed wherever you wish (see Mason's *Woodcraft* for details).

Ninety-four species of birds are reported to eat the fruit which is available the year round, since it does not fall when ripe; rabbits like to eat the young shoots in winter.

Poison-sumac

[*Toxicodendron vernix* (L.) Kuntze.]

Poison-sumac should be known to all who expect to roam through swampy woods, since in wet places (to which it is restricted) it may be common, often forming extensive thickets (occasionally tree size).

It seems of doubtful value to compare the leaves of this species with those of staghorn sumac, since there is so little similarity; however, those of poison sumac (leaflets) are smooth on the margin, not silvery beneath, and have a clear juice which turns black upon drying. Rather, the leaves look like those of white ash except that they are *alternate* and in fall turn a brilliant and attractive red, hence often collected by the unwary. One can be poisoned by bruising or tearing the leaves or twigs of this plant (even in

winter). The subject of poisoning is discussed in more
detail under poison-ivy, and apparently there is little
difference between the two plants in this respect; how-

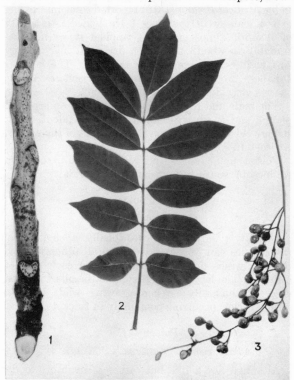

Fig. 122.——Poison-sumac.   Twig × 1¼. Leaf × ½. Fruit
× ½.

ever, some people believe that poison-sumac is worse.

The twigs are stout, yellowish brown, more or less
mottled, with a conical terminal bud, small alternate

laterals, and large, shield-shaped leaf scars. The poisonous juice, as already stated, is clear (not milky) and turns black upon drying. Dr. Bigelow in his *Medical Botany* (1815) reported some interesting experiments with the juice. He boiled a quantity of it until almost resinous, applied it warm (presumably to wood), and when dry a jet black, shiny surface like Japanese lacquer resulted. The familiar Japanese product is made from a poisonous Asiatic species very similar to our poison-sumac.

The male and female flowers are borne on separate plants; and the ivory-colored fruit in pendent, open, stringy clusters remains on the tree most of the winter. At least 16 species of birds eat the fruit.

Poison sumac ranges from New England to southern Minnesota, south to Louisiana and Florida.

Dwarf Sumac

(*Rhus copallina* L.)

Usually a shrub, but occasionally a small tree, this species can be separated from all others by the peculiar wings on the midrib between each pair of leaflets. The leaves are glossy, and the plant is often used ornamentally. When growing wild, it is a scrub species, occurring commonly in thickets.

Poison-Ivy[1]

[*Toxicodendron radicans* (L.) Kuntze.]

Although not a tree but rather a shrub or woody vine, poison-ivy must be mentioned, since it is the most important plant for a woodsman to know—and stay away from. At birth we have some resistance to the poison contained in canals inside the plant. But

---

[1] Poison-oak, a related form with three- to seven-lobed leaflets, hairy beneath, is common in the south.

after the first case of dermatitis we become highly allergic. Therefore try to avoid a first case!

Very little study of the illustrations should be enough to teach anyone how to recognize poison ivy, but perhaps the much repeated jingle should be included.

> Leaflets three quickly flee.[1]
> Berries white (ivory), poisonous sight!

The fruit, technically a drupe (not a berry), persists into the winter, and at least 60 kinds of birds are reported to eat it; hence the common occurrence of poison ivy along hedgerows and in similar places where birds roost.

The twigs are brownish with hairy buds (no scales); the terminal, present; the leaf scars, horseshoe-shaped. The Onondago Indian name for poison-ivy is Ko-hoon-tas, or "stick that makes you sore."

It is usually conceded that actual contact with the plant, especially when one is perspiring, is necessary to produce the eruption. However, it is quite possible to get it from handling such articles as shoes that have crushed the leaves or other parts of the plants, and most certainly one should avoid getting in the smoke from burning ivy.

The poisonous principle, toxicodendrol (or urushiol), one of the phenols, is contained in an oily liquid found in all parts of the plant (except the pollen and perhaps ripe fruit). The poison penetrates the skin soon after contact, and measures, such as washing with strong soap, must be taken almost immediately to be of any value whatsoever. A 10 per cent solution of potassium permanganate is probably the best known and easiest obtained neutralizer, but even this must be applied soon after contact. The poisonous material

---

[1] I have seen only one plant on which some leaflets are also in 5s along with the 3s. The condition has been repeated on this vine for several years.

of poison ivy and poison sumac is exceedingly retentive of its vigor even when separated from the plant, and shoes handled a year or more after they have

FIG. 123.—Poison ivy.   Twig × 1¼.   Leaf × ½.   Fruit × ½.

bruised the foliage or stems, often produce the dermatitis on sensitive persons.[1]

The best way is to recognize the plant and have nothing to do with it.   Except in special cases, only the careless or shiftless get poisoned.

[1] See bulletin, "Poisonivy and Poisonsumac," by W. M. Harlow, New York State College of Forestry, Syracuse, N. Y., 1945.

THE HOLLY FAMILY

Holly

(*Ilex opaca* Ait.)

Holly, for the most part a southern tree, is known to all of us through its use at Christmas for wreaths

Fig. 124.—Foliage and fruit of holly.

and other decorations. Evergreen spiny leaves, red "berries," and a smooth gray bark, marked by warty outgrowths characterize this species. Since male flowers and female flowers are borne on separate trees, it is important to select cuttings from a female tree if one wishes to grow holly for its fruit. When grown from seed, only about one out of ten trees bears fruit; trees begin to flower after 5 to 12 years. Since male trees are necessary for pollination, a few must be provided to ensure fruiting. The "berries" are eaten by at least 18 species of birds; and in this way, holly is apparently extending its range slowly northward. At present, it is found principally on the coastal plain from Massachusetts to Florida and north in the Mississippi Valley to southern Illinois and Indiana.

THE MAPLE FAMILY

THE MAPLES

Leaves.—Opposite, simple (except in one species), usually palmately lobed.

Flowers.—Mostly with perfect and unisexual flowers on the same tree, or male flowers on one tree, the female on another; borne in clusters before, with, or slightly after the leaves unfold.

Fruit.—A typical double, winged fruit which breaks in the center between the seeds.

Twigs.—Moderately slender; buds, opposite; terminal, present; leaf scars, more or less V-shaped.

Remarks.—In general, maple wood is hard and strong; but since this depends upon the kind in question, no more is said here (see individual species). The seeds are eaten by birds, chipmunks, and squirrels, and the twigs browsed in winter by deer and moose.

The eastern maples with simple leaves can be grouped as follows:

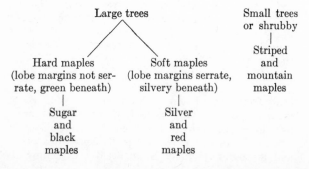

Large trees

Hard maples
(lobe margins not serrate, green beneath)

Sugar
and
black
maples

Soft maples
(lobe margins serrate,
silvery beneath)

Silver
and
red
maples

Small trees
or shrubby

Striped
and
mountain
maples

This leaves out boxelder which has compound leaves (the only species) and wood like that of the soft maples.

KEY TO THE MAPLES

1. Leaves, compound; fruit, V-shaped . . **Boxelder** (p. 246)
1. Leaves, simple and lobed; fruit, more or less U-shaped or spreading **2**

2. Margins of lobes, wavy, not repeatedly toothed. . . . **3**
2. Margins, definitely toothed **5**

3. Leaves, 7-lobed; sap, usually milky (break leaf); two halves of fruit, spread at about 180 deg. (in a straight line) . . . . **Norway maple** (p. 250)
3. Leaves, 5- or 3-lobed; sap, clear; fruit, U-shaped **4**

4. Leaves, 5-lobed, flat, smooth; bark, somewhat roughly platy . . . . **Sugar maple** (p. 239)
4. Leaves, 3-lobed (less commonly 5-lobed), droopy in appearance, more or less densely hairy on the leaf stem and undersurface; bark, more corrugated . . .
Black maple (p. 241)

5. Undersurface of mature leaf, silvery **6**
5. Undersurface of mature leaf, green **7**

6. Leaves, 5-lobed; sides of terminal lobe diverge (V-shaped); fruit, with widely spreading wings; bruised twigs have a rank odor . . . .
Silver maple (p. 241)
6. Leaves, variable, 3- or 5-lobed; terminal lobe, pointed but sometimes like that of silver maple; wings, more nearly parallel; bruised twigs, not offensive . . . . **Red maple** (p. 244)

7. Teeth, coarse and single; bark, brown . . . .
Mountain maple (p. 249)
7. Teeth, fine and double; bark, green with white stripes
Striped maple (p. 248)

Fig. 125.—Sugar maple. *1.* Twig × 1¼. *2.* Flower cluster × ½. *3.* Male flower × 2. *4.* Perfect flower × 2. *5.* Fruit × ¾. *6.* Leaf × ½. *7.* Bark of old-growth tree.

Sugar Maple

(*Acer saccharum* Marsh.)

Appearance.—A medium-sized to large forest tree, 60 to 80 ft. high and about 2 ft. in diameter (max. 135 by 5 ft.), with a short trunk; large, dense oblong to oval crown; and spreading root system. Because of its extreme tolerance to shade, a dense forest is necessary if the tree is to produce a long, clear trunk valuable for lumber.

Leaves.—Opposite, simple; about 4 in. in diameter 5-lobed and with five principal veins; smooth above and below.

Flowers.—Certain trees bear male flowers; others have both male and perfect ones on the same tree. The flowers are yellowish and in a good seed year appear in countless numbers; so much so that at a distance the tree seems to be enveloped in a yellow haze.

Fig. 126.—Open-grown form of sugar maple.

Fruit.—Borne in the fall; U-shaped; wings, about 1 in. long.

Twigs.—Moderately slender, shiny, brown; terminal bud, present, pointed, many scaled; the laterals, opposite; leaf scars, V-shaped.

Bark.—Grayish, on old trees deeply furrowed, irregularly plated or ridged, but variable.

Habitat.—Best growth is made on moist, rich, well-drained soils, but the tree will persist on poorer sites.

Distribution.—The mouth of the St. Lawrence River to southern Manitoba, south to Texas and Louisiana.

FIG. 126a.—Old sugar maple woods with dense young growth beneath.

Remarks.—This is probably the most common and important of the maples. The Indians taught the early French settlers how to make sugar and syrup from the spring sap, an industry that has now become important in one or more of the northeastern states. To taste the cold sweet sap as it drips from the tree is part of every American's birthright! In pioneer days, much potash was exported from the colonies, and sugar maple ash was found to be high in this substance; locally, wood ashes were used in soapmaking. Forges were fired with maple charcoal. As a fuel wood, sugar maple is one of the best and in the campfire yields a bed of hot coals ideal for broiling. The wood is hard and heavy, takes a beautiful polish, and is used for many purposes. Some campers prefer it to spruce for canoe paddles.

Black Maple
(*Acer nigrum* Michx. f.)

This tree is somewhat similar in appearance to sugar maple; and when cut, the wood is not separated from that of the previous species. In typical specimens, the leaves are three-lobed, drooping, definitely velvety below and on the leaf stem; the bark is darker and finely corrugated (not coarsely plated); and the twigs display large warty lenticels. The seed cavity of the fruit is perhaps a trifle larger than in sugar maple. However, intermediate forms are not rare, and some botanists feel that this tree is a variety of sugar maple rather than a distinct species.

Silver Maple
(*Acer saccharinum* L.)

Appearance.—A medium-sized to large forest tree, 60 to 80 ft. high and 2 to 3 ft. in diameter (max. 120 by 5 ft.), with a short trunk, wide-spreading crown, and shallow root system.

Leaves.—Opposite, simple; about 5 in. in diameter, deeply five-lobed, sides of terminal lobe divergent (V-shaped); serrate on the margin; green above; silvery below.

Flowers appear long before the leaves, in early spring, reddish. Some trees bear only male flowers;

Fig. 127.—Black maple. *1.* Twig × 1¼. *2.* Leaf × ½. *3.* Bark.

others both male and perfect ones. In contrast to the flowers of red maple, these have no petals.

Fruit.—Borne in late spring. The seed germinates at once; fruit, the largest of the native maples, with widely divergent wings up to 2 in. long.

Twigs.—Orange-brown to red; terminal bud, present, blunt, with few scales; laterals, opposite; leaf scars, V-shaped; twigs, with a typical disagreeable odor when bruised.

Bark.—At first smooth and gray like that of beech, later breaking up into long, thin, scaly plates curving away at the ends.

Fig. 128.—Silver maple. *1*. Male and perfect flowers respectively × 1. *2*. Fruit. × ¾. *3*. Leaf × ½. *4*. Twig × 1¼. *5*. Bark.

Habitat.—A common swamp or stream-bank tree, but will grow in drier places.

Distribution.—Eastern Canada to South Dakota, south to Arkansas and Florida.

Remarks.—Silver maple has been much used as a shade tree, but the brittleness of the branches is a disadvantage, and it is not a long-lived tree. The wood is softer than that of the hard maples and makes a poorer bed of coals when used in the campfire. Apparently, the bark and also the wood of the two soft maples contain tannic materials (see red maple for details).

Michaux states that a whiter, "more tasty" sugar is made from the soft maples than from sugar maple but that from the same amount of sap, only about one-half as much sugar is obtained.

Red Maple

(*Acer rubrum* L.)

Appearance.—Red maple is a medium-sized forest tree, 50 to 70 ft. high and 1 to 2 ft. in diameter (max. 120 by 4 ft.), with a fairly long trunk, irregularly shaped crown, and a shallow root system.

Leaves.—Simple, opposite, about 4 in. in diameter; typically three-lobed, but sometimes also five-lobed; sides of terminal lobe, convergent or parallel.

Flowers.—Similar to those of silver maple (page 242) but with petals. (Silver maple flowers have none.)

Fruit matures in late spring. The seed germinates at once;[1] fruit, more or less V- or U-shaped, much smaller than that of silver maple; wings, about $\frac{3}{4}$ in. long.

Twigs.—Similar to those of silver maple (page 242) but redder and without a disagreeable odor when bruised.

[1] Sometimes they wait until the following spring.

Fig. 129.—Red maple. *1.* Perfect flowers × ¾. *2.* Male flowers × ¾. *3.* Fruit × ¾. *4.* Leaf × ½. *5.* Bark of old tree (on younger trees similar to that of silver maple). *6.* Flower buds × 1¼. *7.* Twig × 1¼.

Bark.—At first, smooth and gray like that of silver maple; on old trees, entirely different from that on young trees, shaggy or scaly, dark brown.

Habitat.—Like silver maple a typical swamp tree, but also occurs on higher ground.

Distribution.—From southern Newfoundland to Minnesota, south to Texas and Florida.

Remarks.—Like silver maple, the branches are relatively brittle, and decay soon takes place in exposed wounds unless they are treated. The wood is not separated from that of silver maple and as a firewood has similar properties.

At least four authors state that a hot-water extract of the bark can be combined with ferric (iron) or aluminum salts to yield an ink or a dye. Bancroft says that, with aluminous bases, the extract gives a cinnamon color on wool and cotton and, with iron sulfate, a perfect and fast black.

Michaux wrote that hard maple wood can be separated from that of soft maple by a drop of any ferric salt solution on the end grain. Hard maple turns a greenish color, whereas soft maple shows a bright blue. On some 50 samples of each, the author finds that all the hard maple turn greenish, all the soft maple *sapwood* become blue, and about one-half of the heartwood samples tested are also dark blue, the remainder dark greenish blue. The test appears to be permanent, since after 6 months the blocks can still be readily separated by color.

Boxelder   Ash-leaved Maple

(*Acer negundo* L.)

Appearance.—A small to medium-sized tree, 30 to 40 ft. high and 1 to 2 ft. in diameter (max. 75 by 4 ft.), with a short irregular trunk, spreading crown, and shallow root system.

Fig. 130.—Boxelder. *1.* Twig × 1¼. *2.* Leaf × ½. *3.* Fruit
× ¾. *4.* Bark. (*Photograph by R. A. Cockrell.*)

Leaves.—Compound, opposite, with five or seven (less commonly three or nine) leaflets. The leaflets are extremely variable, some deeply lobed or sometimes again compounded.

Flowers.—Male and female borne on separate trees, on slender drooping stems.

Fruit.—A typical maple fruit; characteristically V-shaped, in contrast to that of the other maples.

Twigs—Stout, purplish to greenish brown, often covered with a bluish white bloom; buds, opposite, more or less whitish woolly; the terminal, present, but sometimes poorly developed.

Bark.—Thin, light brown; with narrow rounded ridges; finally more deeply furrowed.

Habitat.—Although doing best on moist sites, this tree is perhaps the hardiest of the maples and will stand great extremes of temperature and dry soils.

Distribution.—From New England to Minnesota, south to Texas and Florida.

Remarks.—This tree has been much planted, especially through the middle west, on account of its hardiness. It is not, however, particularly decorative; it is short lived and of poor form. The wood is similar to that of the soft maples. The fruit furnishes summer and fall food for mice, squirrels, and birds.

Striped Maple   Moosewood

(*Acer pennsylvanicum* L.)

Striped maple is a large shrub or small tree typical of the northern woods. The leaves are large and three-lobed and show small double teeth on the margin. The bark, green with white stripes, is the most distinctive feature. The twigs and buds are smooth, the latter (terminals) covered with a single outer pair of boat-shaped scales.

Michaux mentioned that the twigs and buds were useful as spring browse for horses and cattle. The tree furnishes important winter food (twigs and buds) for deer and moose, and the latter eat the leaves in summer. The buds are also eaten in large numbers by ruffed grouse.

Fig. 131.   Fig. 132.

Fig. 131.—Striped maple. *1.* Leaf × ½. *2.* Buds × 1½.
Fig. 132.—Mountain maple. *1.* Leaf × ½. *2.* Buds × 1½.

Mountain Maple

(*Acer spicatum* Lam.)

This shrub or small tree is an almost constant companion of striped maple in the northern woods or farther south on cool, north-facing slopes. The leaves are usually smaller, either three- or five-lobed, and coarsely singly toothed on the margin. The bark is brownish; the twigs more or less downy; and the terminal bud, with its pair of outer scales, is more conical than that of striped maple.

Mountain maple is browsed along with striped maple, and the buds of both species are important food of the ruffed grouse.

Norway Maple

(*Acer platanoides* L.)

Norway maple is a common, introduced tree similar in general appearance to our native sugar maple. However, the leaves are seven-lobed or veined; the juice is usually milky (break the fresh leaf to see this); the two halves of the winged fruit extend almost in a straight line; and the bark has a closer, smoother pattern. Fruit is often produced in great numbers and carpets the surrounding area. The fruits lie flat all winter; but in spring when the roots begin to grow into the earth, each one raises its flag (wing) to an erect position as if to say "Here am I." It is very amusing to see countless thousands of these wigwagging across an otherwise plain and featureless stretch of lawn. Of course, the first mowing beheads these volunteers, which, it would seem, might otherwise develop very soon into a dense stand of Norway maples.

THE BUCKEYE FAMILY

The Buckeyes

The two species included are small to medium-sized trees found on the west slopes of the Appalachians and through the Ohio Valley. The leaves are opposite, palmately compound, with usually five leaflets, broadest in the middle and tapering both ways. The flowers are in showy clusters, and the capsular fruit splits to release large, shiny, brown seeds ("buckeyes"). Yellow buckeye (*Aesculus octandra* Marsh.) has a fruit relatively smooth on the outside. The fruit of Ohio buckeye (*A. glabra* Willd.) is more

Fig. 133.—Yellow buckeye. *1.* Twig × 1¼. *2.* Leaf × ½. *3.* Fruit × ¾. *4.* Seed × ¾.

spiny, and the twigs and foliage have a disagreeable odor.

The introduced horsechestnut (*A. hippocastanum* L.) is a well-known ornamental tree. The leaflets are usually seven in number, are widest near the apex, and taper narrowly toward the base. The buds, in contrast to those of the native buckeyes, are darker (almost black) and sticky.

Gibson says that the seeds reduced to flour make an excellent library paste—one that will repel all insects on account of its taste. The fresh seeds of these trees are very poisonous to eat, and should never

Fig. 134.—Bark of buckeye. *(Photograph by U. S. Forest Service.)*

be bitten into. It is said that the powdered seeds stirred into a pool will intoxicate the fish and cause them to rise to the surface. This probably violates all game laws (!) but might be used in wilderness survival—if buckeye trees are found.

THE BUCKTHORN FAMILY

Buckthorn

(Rhamnus cathartica L.)

Introduced from Europe, this small tree has now become widely distributed (by birds) throughout the east. The leaves are subopposite (staggered) on the stem, veins parallel margin of leaf (see dogwoods)

Fig. 135.—Flowers and leaves of horsechestnut.

but, unlike the dogwoods, are serrate. The fruit is black, looking something like a small cherry, *but* with a quite different taste. Before the first frosts, it tastes like a cascara pill from which the sugar coating has been removed. After several frosts, the taste is not so bad—a little like that of licorice. As its name indicates, it has cathartic properties; as a matter of fact, the commercial source of cascara sagrada is the bark of a native western tree of this same group, the western cascara tree, or buckthorn.

Fig. 136.—Buckthorn. *1.* Leaf × ½. *2.* Buds × 1½.

Fig. 137.—Basswood. *1.* Leaf × ½. *2.* Bark of old tree, and young sprout, respectively. *3.* Flowers × ½. *4.* Fruits showing persistent leafy bract × ½. *5.* Twig × 1¼.

THE BASSWOOD FAMILY

Basswood

(*Tilia americana* L.)

Appearance.—A medium-sized to large forest tree, 70 to 80 ft. high and 2 to 3 ft. in diameter (max. 125 by 4 ft.), with a well-formed trunk, somewhat rounded crown, and deep, wide-spreading root system. The ability of this species to sprout vigorously accounts for the typical occurrence of the trees in groups—each clump derived from sprouts of a single old-growth tree.

Leaves.—Alternate, simple, more or less circular in outline; about 5 in. in diameter; heart-shaped, often unequally so at the base; serrate on the margin.

Flowers.—Perfect, often with abundant nectar, borne in a cluster, the principal stem of which is attached to a narrow, leaflike blade in turn fastened at the end to the twig.

Fruit.—About $\frac{1}{3}$ in. in diameter, globose, nutlike, with a hard outer shell and edible though small seed. The leaflike blade mentioned above is the most conspicuous feature of the fruit cluster.

Twigs.—Moderately stout, red or green, zigzag; terminal bud, lacking; the laterals, mostly with two visible scales, red, lopsided; in the spring, mucilaginous, edible if one is sufficiently hungry! A prominent botanist once lived on gulls' eggs (none too fresh) and basswood buds for several days when his boat was blown off an uninhabited island in a large lake.

Bark.—At first smooth and gray or dark green; later, narrowly ridged; somewhat scaly or flaky on the surface.

Habitat.—Although best growth is made on moist, deep soils, basswood is found on many other sites and is one of our commonest "hardwoods" (the wood is actually soft).

Distribution.—New Brunswick to Minnesota, south to Kansas, Kentucky, and Delaware.

Remarks.—Basswood is one of our most useful and common forest trees. The wood is soft and ideal for carving. The Iroquois carved false-face masks on a living tree and, after finishing the outside, cut them off to hollow out the back. The wood has a quality of toughness which makes it the chosen material for honey section boxes. Here an extremely thin portion at the corners must stand a 90 deg. angle bend, *dry*, without breaking. Like other soft hardwoods, basswood gives a bright, short-lived campfire with ashes rather than coals remaining.

Fig. 138.—Basswood false face of the Iroquois Indians; carved from a living tree, and then split off.

Basswood bark is perhaps the best source (but not the only one) of woodland rope, string, and thongs or strips for sewing birch bark. Since this tree is such a prolific sprouter, one can usually find plenty of shoots 2 to 6 in. in diameter which can be removed without injuring larger trees. Especially in the spring, the tough bark is easily stripped; and for certain purposes, thongs can be split out without further treatment. To make cordage, the bark is placed in water and weighted down so it will not float. After soaking from two weeks to a month, the soft tissues more or less rot away, leaving the somewhat slippery, fibrous material. The "retting" process can be hurried by pounding, or by simmering the bark in a kettle of water to which has been added a goodly quantity of wood ashes. The long strips are then twisted to form rope or string (see Mason's *Wood-*

craft for details). The Indians claim that this rope is superior to that of the white man, since it is softer on the hands when wet and does not kink when dry. By twisting two strands separately and laying them against each other, a piece of rope of any length can be made, provided one strand is always kept longer than the other, and the new one is twisted in at the short end each time.

The Indians also used the bark fresh from the tree as an emergency bandage for wounds. In Europe, basswoods are called *lindens*, and the bark has been used there for the same purposes and also for weaving coarse cloth, mats, and nets.

Basswood shoots are a staple food for rabbits, and deer also browse the twigs. The buds are eaten by at least four kinds of birds including the pinnated grouse, and the fruit is hoarded by chipmunks.

There are several other species of basswood in this region, but they are so similar that it is doubtful if they should be included in this manual. Even the experts get confused in some cases when identifying them.

THE GINSENG FAMILY

Devil's Walking Stick
Herculesclub

(Aralia spinosa L.)

Fig. 139.—Subleaflet of twice-compound leaf from devil's walking stick. × ½.

Devil's walking stick is a large shrub or small tree with extremely stout, spiny twigs and twice-compound leaves 2 to 4 ft. long. It is native in the southern part of the region covered by this book and is commonly planted farther north as an "ornamental." It is, however, more grotesque than ornamental.

THE TUPELO FAMILY

Black Tupelo   Blackgum   Sourgum

(*Nyssa sylvatica* Marsh.)

Appearance.—A medium-sized tree, 40 to 60 ft. high and 1 to 2 ft. in diameter (max. 125 by 5 ft.),

Fig. 140.—Black tupelo. *1.* Twig 1¼. *2.* Leaf × ½. *3.* Fruit × ¾. *4.* Bark.

with an irregularly shaped, often flat-topped, or conical crown; long trunk; and shallow root system.

Leaves.—Alternate, simple; about 4 in. long; shiny; widest above the middle; entire on the margin.

Flowers.—Inconspicuous, greenish; the sexes separated on different trees; certain individuals also with perfect flowers.

Fruit.—A blue drupe about ⅓ in. in diameter, borne on a long slender stem, singly, in 2s or in 3s. The pit is ridged, one of the best features for recognizing it.

Twigs.—Slender to moderately stout; short spur shoots, usually present; pith, when cut lengthwise, shows faint bars of darker tissue crossing at intervals; terminal bud, present; the laterals, alternate; leaf scars, conspicuous, each with three large bundle scars.

Bark.—On mature trees deeply and narrowly fissured, with oblong blocks, resembling alligator leather.

Habitat.—This is a typical swamp-forest tree, found as well along lake shores, but growing on drier sites when planted.

Distribution.—From Maine to Missouri, south to Texas and Florida.

Remarks.—When first seen from a distance, a small tree of this species with its shiny leaves and dwarfed branchlets looks like a pear tree. In the fall, the leaves of black tupelo are ablaze with scarlet, thus making the tree stand out from some of its more somber neighbors.

The wood is noted for its extreme toughness and difficulty in splitting; probably for this reason, it was used for making pipes that carried salt water to the salt works at Syracuse, N. Y., in early colonial days. The ends of the log pipes could be fitted together without restraining iron bands (which would soon rust).

THE DOGWOOD FAMILY

KEY TO THE DOGWOODS[1]

1. Leaves, alternate . . **Alternate-leaved dogwood** (p. 263)
1. Leaves, opposite **2**

[1] Although shrubby with two exceptions, they are included because they are so common.

2. Twigs, purplish with a white bloom, somewhat diamond-shaped in cross section; fruit, bright red, in compact heads; tree, with platy or blocky bark.
Flowering dogwood (p. 260)
2. Twigs, red, green, orange, or gray, circular; fruit, white or blue, shrubby........................ 3

3. Twigs, bright red or greenish...................... **4**
3. Twigs, orange- or reddish brown or gray........... **6**

4. Twigs, with many conspicuous purple spots; fruit, white............**Round-leaved dogwood** (p. 263)
4. Twigs, lacking purple spots; fruit, white or blue.... **5**

5. Leaves, broad; twigs, smooth; pith, white...........
Red-stemmed dogwood (p. 263)
5. Leaves, narrow; twigs, silky; pith, brown............
Silky dogwood (p. 263)

6. Twigs, reddish brown; leaves, more or less rough above............**Rough-leaved dogwood** (p. 263)
6. Twigs, grayish on older growth; leaves, smooth....
Panicled dogwood (p. 263)

Flowering Dogwood

(*Cornus florida* L.)

Appearance.—A small tree or large shrub, 10 to 30 ft. high and about 6 in. in diameter (max. 40 by 1½ ft.), with a short, clear trunk and bushy, somewhat layered-appearing crown.

Leaves.—Opposite, simple; about 3½ in. long; with an entire margin and peculiar type of venation called *arcuate* (secondary veins parallel the margin) found in all the dogwoods and rare in other groups of woody plants.

Flowers.—Small, inconspicuous, perfect; forming a flat-topped cluster or head. The four large, petal-

FIG. 141.—Flowering dogwood. *1.* Twig × 1¼. *2.* Flower bud × 1¼. *3.* Flower cluster showing the white bracts often mistaken for petals × ½. *4.* Flower × 3. *5.* Cluster of fruits × 1. *6.* Leaf × ½. *7.* Flower clusters. (*Photograph by C. A. Brown.*) *8.* Bark.

like bracts (actually expanded bud scales) are very striking and lead most people to believe that the entire cluster is a single flower.

Fruit.—A bright red drupe, about $\frac{1}{3}$ in. long, borne in compact clusters.

Twigs.—Slender, purplish; angled; more or less covered with a whitish bloom; terminal bud, present; the scales meet in a straight line without overlapping; laterals, opposite; flower buds look like miniature urns.

Bark.—Broken by deep fissures into small blocks, the whole looking like alligator hide.

Habitat.—Mostly as an understory tree in moist hardwood forests.

Distribution.—Maine to Kansas, south to Texas and Florida.

Remarks.—This is the only dogwood important for its wood (hard and heavy, something like that of hard maple), and it is also the one most widely planted as an ornamental. It can be grown from seed, which requires 2 years to germinate, and under favorable conditions begins to flower after 10 or 12 years.

Fig. 142.—Alternate-leaved dogwood *1.* L e a f × ½. *2.* Bud × 1.

The inner bark is bitter and was used as a substitute for quinine. From the bark of the smaller roots, the Indians made a red dye. The split ends of small branchlets can be used as a toothbrush; it is said they whiten the teeth exceptionally.

At least 86 species of birds, including the ruffed grouse, bobwhite, and turkey, eat the fruit, which is also sought by squirrels. In this way, the tree furnishes food for wild life well into December.

Alternate-leaved Dogwood

(*Cornus alternifolia* L.)

This is also a small tree like the flowering dogwood but is commonly not so large. The leaves are similar, but narrower and alternate (the only alternate-leaved dogwood). The fruit is blue and furnishes food for a reported 11 species of birds. The twigs, in contrast to those of flowering dogwood, are greenish.

The other common dogwoods are shrubs, but so widespread that their twig features are included.

Species	Twigs
Cornus racemosa (panicled)...	Slender, after the first year grayish
C. stolonifera (red-stemmed)..	Red, smooth, with white pith
C. amomum (silky)...........	Dark red, silky, with brown pith
C. rugosa (round-leaved).....	Green to red, with purple spots
C. asperifolia rough-léaved....	Reddish brown, more or less rough, with brown pith

THE HEATH FAMILY

Great Rhododendron

(*Rhododendron maximum* L.)

Mostly shrubby and occurring in thickets, this species is occasionally a small tree. The showy flowers and glossy evergreen leaves make it a favorite ornamental. However, an acid soil is necessary for successful growth.

Mountain-laurel

(*Kalmia latifolia* L.)

This is one of our most beautiful native shrubs, or occasionally small trees, with its shiny evergreen leaves (narrower and more pointed than those of rhododendron) and great masses of white or pinkish bloom. Like the great rhododendron, mountain-laurel is for the most part found in the Appalachians and adjacent regions, on acid or peaty soils.

FIG. 143.—Great rhododendron. *1.* Flower bud × 1. *2.* Fruits × ¾. *3.* Flowers × ½. (*Photograph by D. M. Brown.*)

Sourwood

[*Oxydendrum arboreum* (L.) DC.]

This is a small to medium-sized tree ranging from southern Indiana and Pennsylvania southward. The leaves are alternate, deciduous, finely toothed, and sour tasting. The flowers are bell-shaped, looking something like those of blueberry. Like others in this family, sourwood is rarely, if ever, found on limy soils.

THE EBONY FAMILY

Persimmon

(*Diospyros virginiana* L.)

Persimmon, for the most part a southern tree, comes as far north on the coast as southern Connecticut and passes westward through central Pennsylvania and the Ohio Valley states to southeastern Iowa.

The leaves are alternate, simple, and somewhat oblong, with smooth margins; the fruit is the well-

FIG. 144.—Persimmon. *1.* Bark. *2.* Seed × ¾. *3.* Fruit × ¾. *4.* Leaf × ½. *5.* Buds × 3.

known persimmon, a large berry with several large, flattened seeds. Six birds including the bobwhite and animals such as the raccoon, spotted skunk, fox, and opossum eat the fruit.

In 1539, De Soto's companions learned of the fruit from the Indians. It is written that "they traveled seven daies journie through a desert, and returned eating greene plums and stalkes of mais."

THE OLIVE FAMILY[1]

THE ASHES

Leaves.—Opposite; compound; leaflets, more or less elliptical, serrate or entire.

Flowers.—(1) Perfect, (2) male and female on separate trees, or (3) perfect and unisexual on the

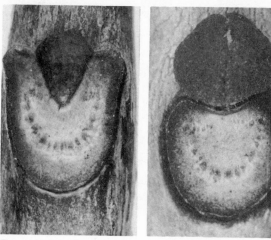

Fig. 144a.—Leaf scars of white ash (left), and black ash (right) showing differences in structure × 8.

same tree; small, the male in crowded clusters, the female in more open stringy ones (page 268).

Fruit.—Terminally winged; in several species, not unlike the blade of a canoe paddle in shape.

Twigs.—Stout; buds, opposite; the terminal, present; leaf scars, large and conspicuous, slightly protruding in such a way that even at some distance,

[1] Fringetree (*Chionanthus virginica L.*), mostly a small southern species, has simple, opposite leaves, and a blue drupe for a fruit.

they can be seen like paired steps along the twig (giving a jointed appearance).

Remarks.—In general, the ashes produce hard, strong wood (black ash excepted). In the campfire, ash wood burns to a bed of hot coals ideal for broiling. The seeds are eaten by a few birds, and deer browse the winter twigs. An early English writer (Evelyn) said that the inner bark of ash was used to write upon before the advent of paper, and perhaps the native ashes could be used similarly.

KEY TO THE ASHES

1. Twigs, 4-sided on account of corky ridges; sap turns blue on exposure.................**Blue ash** (p. 273)
1. Twigs, circular in cross section; sap, not blue........ **2**

    2. Leaflets, lacking a stem, joined directly to the midrib; bark, gray, not showing a diamond pattern; seed cavity of fruit, indistinct...**Black ash** (p. 271)
    2. Leaflets have a distinct stem; bark, with interlacing ridges; seed cavity, distinct..................... **3**

3. Leaflets, sparingly toothed, elliptical to almost oval...
                              White ash (p. 267)
3. Leaflets, conspicuously toothed, narrow, elliptical to lance-shaped................................ **4**

    4. Twigs and leaves, more or less hairy or velvety....
       Now considered one species.   **Red ash** (p. 269)
    4. Twigs and leaves, smooth......**Green ash** (p. 270)

White Ash
(*Fraxinus americana* L.)

Appearance.—A medium-sized to large forest tree, 70 to 80 ft. high and 2 to 3 ft. in diameter (max. 125 by 6 ft.), with a long, straight trunk; irregularly shaped, rather open crown; and a deep, wide-spreading root system.

Leaves.—Opposite, pinnately compound, with usually 7 (rarely 3 to 11) oval to elliptical leaflets which

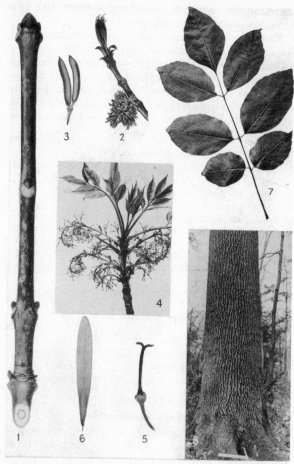

Fig. 145.—White ash. *1*. Twig × 1¼. *2*. Clusters of male flowers × ½. *3*. Male flower × 4. *4*. Clusters of female flowers × ½. *5*. Female flower × 4. *6*. Fruit × 1. *7*. Leaf × ⅓. *8*. Bark.

are about 4 in. long, sparingly toothed or smooth on the margin, and attached by a stem to the midrib.

Flowers.—Small and inconspicuous; male and female borne on separate trees.

Fruit.—Terminally winged, resembling a canoe paddle in shape, the seed cavity at the base.

Twigs.—Stout, grayish or greenish brown; terminal bud, present, rounded; the first pair of lateral buds, situated directly below the terminal; leaf scars, conspicuous, deeply notched on the upper edge.

Bark.—At first smooth, often with an orange cast, soon finely furrowed and showing a diamond-shaped pattern.

Habitat.—Although making best growth on deep, moist soils, white ash is the most common and widespread of the ashes and is found in many different hardwood mixtures.

Distribution.—From Nova Scotia to southern Minnesota, south to eastern Texas and Florida.

Remarks.—This is the most important of the native ashes and a valued member of our hardwood forest. The wood is hard and strong and used for many purposes, especially tool handles of every description. It burns to a bed of hot coals suitable for broiling.

Michaux says that a leaf of white ash rubbed on a mosquito bite or bee sting relieves the itching at once! This is certainly worth trying.

Red Ash[1]

(*Fraxinus pennsylvanica* Marsh.)

Red ash is a swamp or stream-bank tree whose bark is similar to that of white ash. The leaflets are narrower and usually more toothed, and the leaf scars nearly straight on their upper edges. The best feature of identification is the wool or velvet

[1]This is no longer separated from green ash, page 270.

that more or less covers the new twigs, the framework of the leaf, and the undersurface of the leaflets. The fruits are narrower than those of white ash, and

Fig. 146.—Red ash. *1.* Twig × 1¼. *2.* Leaf × ⅓. *3.* Fruit × 1.

the wing is poorly developed or lacking along the seed. (Red ash ranges from Nova Scotia to Manitoba and Kansas, south to Missouri, Mississippi, and Georgia.)

Green-ash[1]

[*Fraxinus pennsylvanica* var. *lanceolata* (Bork.) Sarg.]

When an otherwise typical red ash is found with no velvet or hair on the twigs or leaves, it then

[1]No longer recognized as separate from *Fraxinus pennsylvanica*, page 269. Green ash is the preferred name, and "red ash" goes out!

becomes known as *green ash!* Otherwise there seems to be little difference between the two forms. However, green ash covers a much wider range, which extends to and in a few places beyond the Rocky Mountains.

Fig. 147.—Green ash. *1*. Twig × 1¼. *2*. Cluster of fruits × ⅓.

Black Ash

(*Fraxinus nigra* Marsh.)

Appearance.—A medium-sized tree, 40 to 50 ft. high and about 1½ ft. in diameter (max. 90 by 4 ft.), with an often crooked but clear trunk; open, scraggly crown; and shallow root system.

Leaves.—Opposite, pinnately compound, with about 9 (7 to 13) oblong leaflets; serrate on the margin and attached directly to the midrib.

Flowers.—Male and perfect flowers on the same or different trees, small, inconspicuous.

Fruit.—With a broad wing and indistinct seed portion.

Twigs.—Stout, gray; terminal bud, present, conical; the first pair of lateral buds placed some distance below the terminal; leaf scars, straight on their upper edge, otherwise somewhat vertically oval-shaped.

Fig. 148.—Black ash. *1.* Twig × 1¼. *2.* Leaf × ⅓. *3.* Fruit × 1. *4.* Bark.

Fig. 149.—Making a black ash packbasket.

Bark.—Quite different from that of the preceding ashes; at first, smooth and gray; later, shallowly ridged or flaky; never with the diamond-shaped areas found in white, red, and green ash.

Habitat.—A typical swamp or stream-bank tree never found on dry soils; if found on a hillside, look for a spring hole or near-by depression.

Distribution.—From Newfoundland to Lake Winnipeg, south to the Ohio Valley and Delaware.

Remarks.—The wood of this tree is inferior in strength to that of the other ashes, but from early times it has had a use all its own[1] —for baskets, especially pack baskets. The northern Indians by pounding a log with wooden clubs caused the wood to separate in thin slats, each one the thickness of a year's growth. Probably a better method is to chop or otherwise work out a piece rectangular in cross section with the rings exactly at right angles to the longest dimension. Then by pounding, one will get slats all the same width—and much more easily.

Fig. 150.—Blue ash. *1*. Twig showing corky wings × 1¼. *2*. Fruit × 1.

Blue Ash

(*Fraxinus quadrangulata* Michx.)

Blue ash is for the most part an Ohio Valley and upper Mississippi Valley tree, similar in its fruit and

[1] More recently other ashes and oak have been used.

bark to black ash. The only features necessary for identification are the peculiar corky growths on the twig, giving it a four-sided appearance, and the sap which turns blue when exposed to the air. The pioneers made a blue dye by macerating the bark in water. The wood of blue ash is like that of white ash and is used for the same purposes. This tree is common, especially on dry limestone uplands.

THE TRUMPET CREEPER FAMILY

Hardy Catalpa

(*Catalpa speciosa* Ward.)

This species is a small tree originally native to a restricted area in the Mississippi Valley but now spread by planting through the entire east. The large heart-shaped leaves occur, especially on young trees, in whorls of three or more, rarely opposite. The showy flowers and later the long so-called *Indian beans* (really capsules), about ½ in. in diameter, make the tree attractive as an ornamental. The seeds, many in each capsule, are flattened, with a wing bearded at each end. The twigs with large, cupped leaf scars are also distinctive. The soft brown wood is very durable.

Common Catalpa

(*Catalpa bignonioides* Walt.)

So-called *common catalpa* is similar to hardy catalpa except that the flowers are purple spotted and the fruit is more slender with thinner walls. Sargent mentions that nectar glands occur in the vein axils of the leaves, and H. P. Brown has observed the same thing in hardy catalpa.

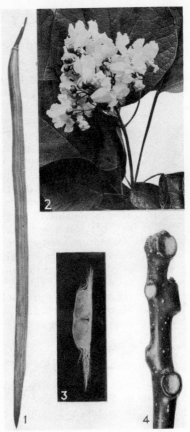

FIG. 151.—Catalpa. *1.* Fruit $\times \frac{1}{3}$. *2.* Flowers $\times \frac{1}{2}$. *3.* Seed
$\times \frac{3}{4}$. *4.* Twig $\times 1\frac{1}{4}$.

THE HONEYSUCKLE FAMILY

Nannyberry

(*Viburnum lentago* L.)

This large shrub or small tree is for the most part a weed, at least to anyone who has ever tried to get rid of a patch of it! When cut down, it sprouts in a dozen new places. The opposite, serrate leaves are more or less broadly elliptical and are borne on peculiar grooved leaf stems which usually have a small edge of wrinkled leaf tissue along each side. The buds are gray, about ¾ in. long, lance-shaped; and the fruit is a blue-black drupe persisting throughout the winter, hence available to birds of which at least the ruffed grouse take advantage. Nannyberry ranges from eastern Canada to Manitoba, southward to Mississippi and Georgia.

Fig. 152.—Nannyberry. *1.* Leaf × ½. *2.* Bud × 1.

Black Haw

(*Viburnum prunifolium* L.)

Similar to the preceding species, black haw has smaller leaves (about 2 in. long) less often winged, twigs with spine like spurs, and shorter buds with rusty hairy scales.

Selected References

BAILEY, L. H. The Cultivated Conifers. The Macmillan Company, New York. 1934.

BERRY, E. W. Tree Ancestors. The Williams & Wilkins Company, Baltimore. 1923.

BLAKESLEE, A. F., and C. D. JARVIS. Trees in Winter. The Macmillan Company, New York. 1916.

BROWN, H. P. Trees of the Northeastern States. The Christopher Publishing House, Boston. 1937.

CANADIAN DOMINION FOREST SERVICE. Native Trees of Canada. Kings Printer, Ottawa. 1949.

CLUTE, W. N. The Common Names of Plants and Their Meanings. W. N. Clute and Co., Indianapolis. 1931.

DEAM, C. C. Trees of Indiana. Department of Conservation Indiana Publication, 13, 1921.

EMERSON, G. B. Trees and Shrubs of Massachusetts. 1846.

FERNALD, M. L. Edible Wild Plants of Eastern North America. Idlewild Press. 1943.

FISHER, P. L., A. H. BRIGGS, and others. Game Food and Cover Plants of the Lake States. Lake States Forest Experiment Station. 1935.

GATES, F. C., and others. Trees in Kansas, Vol. 47, No. 186-A. Kansas State Board of Agriculture, Topeka. 1928.

GIBSON, H. H. American Forest Trees. Hardwood Record, Chicago. 1913.

HARLOW, W. M., and E. S. HARRAR. Textbook of Dendrology. McGraw-Hill Book Company, Inc., New York. 3rd ed. 1950.

HARRINGTON, H. D. The Woody Plants of Iowa in Winter Condition, p. 276. University of Iowa. 1934.

HOUGH, R. B. Handbook of the Trees of the Northern States and Canada. Lowville, New York. 1907.

ILLICK, J. S. Pennsylvania Trees. *Pennsylvania Department of Forests and Waters, Bulletin* 11 (reprinted), Harrisburg. 1928.

———. Tree Habits: How to Know the Hardwoods. American Naturalist Association, Washington. 1924.

KEPHART, H. Camping and Woodcraft. The Macmillan Company. New York. 1921.

MASON, B. S. Woodcraft. A. S. Barnes & Company, New York. 1939.

MICHAUX, F. A. North American Sylva. Vols. 1, 2, and 3. D. Rice and A. N. Hart, Philadelphia. 1857.

MILLER, R. B., and L. R. TEHON. The Native and Naturalized Trees of Illinois. Department of Education, Urbana, Ill. 1929.

MORTON, B. R. Native Trees of Canada. Forest Branch, *Department of the Interior, Bulletin* 61, Ottawa. 1917.

MUENSCHER, W. C. Keys to Woody Plants. Ithaca, N. Y. 1922.

——— Poisonous Plants of the United States. The Macmillan Company, New York. 1939.

NUTTALL, T. North American Sylva. Vols. 1 and 2. D. Rice and A. N. Hart, Philadelphia. 1859.

OTIS, C. H. Michigan Trees. University of Michigan, Ann Arbor. 1926.

PERRY, G. S. The Common Trees and Shrubs of Pennsylvania. *Department of Forests and Waters, Bulletin* 33. Harrisburg, Pa.

REHDER, A. Manual of Cultivated Trees and Shrubs. 2d ed. The Macmillan Company, New York. 1940.

ROGERS, W. E. Tree Flowers of Forest, Park, and Street. Published by the author, Appleton, Wis. 1935.

ROSENDAHL, C. O., and F. K. BUTTERS. Trees and Shrubs of Minnesota. University of Minnesota Press. 1928.

SARGENT, C. S. Manual of the Trees of North America. Houghton Mifflin Company, Boston. 1926.

SETON, E. T. The Forester's Manual. Doubleday, Doran & Company, Inc., New York. 1912.

TRELEASE, W. Winter Botany. Urbana, Ill. 1918.

VAN DERSALL, W. R. Native Woody Plants of the United States: Their Erosion-control and Wildlife Values. *U.S. Department of Agriculture, Miscellaneous Publication*, 303. 1938.

WERTHNER, W. B. Some American Trees. The Macmillan Company, New York. 1935.

WHITE, J. H. The Forest Trees of Ontario. King's Printer, Toronto. 1925.

INDEX

WOODCRAFT AND CAMPING

by George W. Sears ("Nessmuk")

Written at a time when woodlore and woodcraft were vital skills, and when America's wilderness regions offered a true opportunity for "roughing it," this book has remained a classic through three or four generations of readers. The author, George W. Sears, is best known as "Nessmuk," and if ever there was a Daniel Boone, this must have been the man! His knowledge of how to get along on camping, hiking, and hunting trips—and of how to get the most out of such experiences—is unsurpassed.

No book has ever inspired so many readers to get out and try out-door living for themselves. Nessmuk is so sincerely appreciative of the pleasures and peace of mind that come from direct contact with nature that the appeal of the woods becomes almost irresistible as you read through his pages. A treasure-chest of useful, specific information, instructions, and suggestions on every aspect of woodcraft, his book is also an inspiration that cannot fail to arouse genuine enthusiasm among Scouts, day-school students, young campers, and other youth groups. And adult readers find its appeal just as exciting.

The author's straightforward devotion to out-door life and recreation provides refreshing reading whether you are a nature enthusiast or not. But it is to the confirmed outdoorsman, to the reader in charge of group nature activity, and to the novice camper or hiker that the book is primarily directed. With its fund of practical advice and its contagious spirit, it should be a constant companion on trips and woodland outings, just as it has been for countless other readers for over fifty years.

Slightly abridged and altered republication of 1920 edition. Index. 12 illustrations. ix + 105pp. 5⅜ x 8½.   21145-2 Paperbound **$1.25**

LIFE HISTORIES OF NORTH AMERICAN BIRDS OF PREY

by Arthur Cleveland Bent

The all-inclusiveness of Bent's volumes on North American birds has made them classics of our time. Arthur Cleveland Bent was one of America's outstanding ornithologists, and his twenty-volume series on American birds, published under the auspices of the Smithsonian Institution, forms the most comprehensive, most complete, and most-used single source of information in existence. No ornithologist, conservationist, amateur naturalist or birdwatcher should be without a copy; yet copies are increasingly hard to come by. Now, however, Dover Publications is republishing at inexpensive prices the entire series, beginning with the two-volume set on birds of prey.

In these two volumes, the reader will find an encyclopedic collection of information about more than a hundred different subspecies of hawks, eagles, falcons, buzzards, condors, and owls. Not a group of general descriptions, but a collection of detailed, specific observations of individual flocks throughout the country, it describes in readable language and copious detail the nesting habits, plumage, egg form and distribution, food, behavior, field marks, voice, enemies, winter habits, range, courtship procedures, molting information, and migratory habits of every North American bird of prey from the ubiquitous barn owl to the extinct caracara of Guadaloupe Island.

Completely modern in its approach, the study was made with full recognition of the difficulties inherent in the observation and interpretation of wild life behavior. For that reason, not only the reports of hundreds of contemporary observers throughout the country were utilized, but also the writings of America's great naturalists—Audubon, Burroughs, William Brewster—of the past. The complete textual coverage is supplemented by 197 full-page plates containing close to 400 photographs of nesting sites, eggs, and the young of important species at various stages in their growth.

Unabridged republication of 1st (1937, 1938) edition. Index for each volume. Bibliographies of 403, 520 items. 197 full-page plates. Total of 907pp. 5⅜ x 8½.

Vol. I:	T931 Paperbound	**$4.00**
Vol. II:	T932 Paperbound	**$4.00**
The set:	T931-2 Paperbound	**$8.00**

HOW TO KNOW THE WILDFLOWERS
by Mrs. William Starr Dana

This well-known classic of nature lore has introduced hundreds of thousands of readers to the wonder and beauty of the wild flowers of most of the United States and Canada. Written with grace and charm, it is not only the handiest field guide to wild flowers, it is also a most pleasant and delightful book, packed full of interesting lore about plants and flowers.

To enable the reader to identify any given flower as easily as possible, Mrs. Dana has first classified plants by the color of their typical flowers: white, red, green, blue, yellow, pink, etc. She has then arranged the flowers within each color group according to their time of blossoming. As a result, if you should come upon a blue flower in June, you can turn instantly to the blue flowers, early section, and see a clear illustration of the plant you want to find—even if you have never had any botanical training at all. This combination of color and time classification makes this the easiest wild flower guide to use.

Mrs. Dana's coverage of the wild flowers of the Eastern and Central United States and Canada is very thorough, all in all more than 1,000 important flowering, berry-bearing and foliage plants. More than 170 full-page plates illustrate the most important and most typical plants (showing foliage, flower, growth habit, roots, fruit, and whatever else is needed), so that you can identify your find at a glance. These illustrations by Marion Satterlee are famous as being both the most realistic and most interesting of modern floral drawings. Many readers have colored them as they have found individual flowers, thus keeping a permanent record of their field trips.

A full text provides you with complete botanical information about each important plant, information about the history, uses, folklore, habitat and other material for each plant, while introductory chapters explain principles of botanical classification and description for those interested.

Unabridged reproduction of enlarged (1900) edition, with additional illustrations. Nomenclature modernized by Clarence J. Highlander. 174 full-page illustrations, more than 150 figures. xlii + 438pp. 5⅜ x 8½.   20332-8 Paperbound **$2.75**

WESTERN FOREST TREES
By James B. Berry

For years this work has been a standard guide to the trees of the Western United States. Anyone who enjoys nature walks or hikes will want to own a copy of this woodlore classic, as will campers, vacationers, and all lovers of the outdoors.

This handy manual covers over 70 different subspecies of trees, ranging from the Pacific shores to the Rocky Mountain forests—as far east as Western South Dakota. All pertinent information is supplied for each type of tree: range, occurrence, growth habits, appearance and particularities of leaves, bark, fruit, twigs, etc., its wood, distinguishing features, and uses.

The book is divided into sections based on leaf characteristics: trees with needle-like leaves (pine, fir, spruce, redwood, hemlock, larch); trees with scale-like leaves (cedar, cypress, juniper); compound broadleaf trees (walnut, ash, locust, mesquite, California buckeye); lobed or divided broadleaf trees (maple, oak, sycamore); and simple broadleaf trees—including leaves with toothed margins (poplar, birch, alder, cherry, black willow) and with entire margins (desert willow, mahogany, gum, dogwood, laurel, madroña). This arrangement, together with the nearly 100 accompanying illustrations (mostly full-size) of buds, branches, twigs, leaves, and the like, provides an easily-used identification key covering virtually every tree of the area.

A long introductory section explains proper procedures in tree and wood identification and describes the general properties and structure of the various woods. An analytical key to porous and non-porous woods is also furnished, as are several subsidiary fruit, bark, leaf, and twig keys within the text for helpful reference purposes. All persons interested in the trees of the American West will find this little handbook just the thing to help them increase their knowledge and appreciation of their favorite woodlands.

Revised edition. Preface. 12 photographs. 85 fine line illustrations by Mary E Eaton. Index. xii + 212pp. 5⅜ x 8 .   21138-X Paperbound **$3.00**

MANUAL OF THE TREES OF NORTH AMERICA

by Charles Sprague Sargent

The greatest dendrologist America has ever produced was without doubt Charles Sprague Sargent, Professor of Arboriculture at Harvard and Director of the Arnold Arboretum in Boston until his death in 1927. His monumental "Manual of the Trees of North America," incorporating the results of 44 years of original research, is still unsurpassed as the most comprehensive and reliable volume on the subject. Almost every other book on American trees is selective, but this one assures you of identifying any native tree; it includes 185 genera and 717 species of trees (and many shrubs) found in the United States, Canada, and Alaska. 783 sharp, clear line drawings illustrate leaves, flowers, and fruit.

First, a 6-page synoptic key breaks trees down into 66 different families; then, an unusually useful 11-page analytical key to genera helps the beginner locate any tree readily by its leaf characteristics. Within the text over 100 further keys aid in identification. The body of the work is a species by species description of leaves, flowers, fruit, winterbuds, bark, wood, growth habits, etc., extraordinary in its fullness and wealth of exact, specific detail. Distinguishing features of this book are its extremely precise locations and distributions; flower and leaf descriptions that indicate immaturity variations; and a strong discussion of varieties and local variants.

Additional useful features are a glossary of technical terms; a system of letter keys classifying trees by regions; and a detailed index of both technical and common names (index, glossary, and introductory keys are printed in both volumes.) Students and teachers of botany and forestry, naturalists, conservationists, and all nature lovers will find this set an unmatched lifetime reference source. "Still the best work," Carl Rogers in "The Tree Book."

Unabridged and unaltered reprint of the 2nd enlarged 1926 edition. Synopsis of Families. Analytical Key of Genera. Glossary. Index. 783 illustrations, 1 map. Total of 982pp. 5⅜ x 8.

T277 Vol I Paperbound **$3.00**
T278 Vol II Paperbound **$3.00**
The set **$6.00**

GUIDE TO SOUTHERN TREES

by Ellwood S. Harrar

Dean of the School of Forestry, Duke University

and Dr. J. G. Harrar

President, Rockefeller Foundation

On nature walks, on hikes, while camping out, or even while you're driving through a wooded area, this 700-page manual will be your unfailing guide. With it, you'll be able to recognize any one of more than 350 different kinds of trees, from the common pine, cypress, walnut, beech, and elm to such seldom-seen species as Franklinia (one of the world's rarest trees, last seen growing naturally in 1790).

"Guide to Southern Trees" covers the entire area south of the Mason-Dixon line from the Atlantic Ocean to the Florida Keys and western Texas. An astonishing amount of information is packed into the description of each tree: habit, leaves, flowers, twigs, bark, fruit, habitat, distribution, and importance, as well as information of historical or commercial significance. Conifers and broadleaved trees are both fully covered, in readable and non-technical language—an especially helpful feature for the beginner and the amateur.

There are 200 full-page delineations (primarily of leaf structure) all carefully drawn to provide the maximum amount of precise, detailed information necessary for identification purposes. In addition, there is a 20-page synoptic key to the generic groups, which will help you find what family a particular tree belongs to, and finding keys for each family as well. Thus, you can use just two keys to find any unfamiliar tree in a matter of minutes. Finally there is a full explanatory introduction covering nomenclature, classification procedures, and important botanical functions for the layman.

The features listed above make this perhaps the most comprehensive guide available at such an inexpensive price. Amateur naturalists, teachers of natural science, Scout Masters, camp counselors, foresters, botanists, conservationists, gardeners, hikers, hunters, and everyone concerned with and interested in trees, from beginner to expert, will find this book an indispensable companion.

Unabridged republication of 1st (1946) edition. Index. 81-item bibliography. Glossary. 200 full-page illustrations. x + 712pp. 4½ x 6½.   T945 Paperbound **$3.50**

HOW TO BECOME EXTINCT

by Will Cuppy

> "The last two Great Auks in the world were killed June 4, 1844, on the island of Eldey, off the coast of Iceland. The last Passenger Pigeon, an old female named Martha, died September 1, 1914, peacefully, at the Cincinnati Zoo. I became extinct on August 23, 1934. I forget where I was at the time, but I shall always remember the date."

So wrote Will Cuppy, chronicler of "The Decline and Fall of Just About Everybody" and author of nearly a dozen other classics of American humor, including "How to Attract the Wombat" (Dover, $1.00). In this collection, Cuppy discusses the extinction of the dinosaur, the plesiosaur, the pterodactyl, the wooly mammoth, the dodo, and the giant ground sloth, and does a pretty good job on quite a few other, less extinct fish and reptiles.

The result is a deliriously funny anthology of 40 short pieces, each stamped with the unmistakable and indelible style of a master humorist. "Do Fish Think, Really?," "Note on Baron Cuvier," "Fish Out of Water," "Own Your Own Snake," "Aristotle, Indeed!," and an appendix containing "Are the Insects Winning?" and "Thoughts on the Ermine" are some of the longer essays interspersed among Cuppy's wry observations, sassy descriptions, and downright disrespectful comments on the cod (who has no vices but whose virtues are awful), the perch ("You have the mind of a perch" is the worst thing you can say to a fish), the tortoise (who is slow, plodding, herbivorous, and against all modern improvements), the boa constrictor, the cobra, and others.

The stamp of the author's style, which kept America amused for two decades, is not the only thing that is indelible about this book. The humor, too, defies the years and comes down to us, in this first reprint in 25 years, still fresh and still very much funnier than practically anyone else has ever been in writing. Complementing Cuppy's text are 51 line drawings by the incomparable William Steig, who is still among our most popular cartoonists.

Complete, unabridged republication of original (1941) edition. 51 illustrations by William Steig. x + 106pp. 5⅜ x 8½.   21273-4 Paperbound **$1.00**

THE STORY OF GARDENING:

From the Hanging Gardens of Babylon to the Hanging Gardens of New York

by Richardson Wright

The author of this book was the editor of "House and Garden Magazine" for 35 years and a member of the board of directors of New York's Horticultural Society. He was also one of the most widely-read writers on gardening, for his love of the subject shows through every page of his books. This warm and informative text, one of Wright's best, covers 6,000 years of gardening history, from the earliest efforts of primitive men to the gardens high atop New York's skyscrapers.

There are discussions of Chinese and Japanese gardens; the formal Mohammedan gardens with their water pools; early Greek and Roman gardens and their statuary; Italy's monastic and villa gardens; Spanish patios, the source of present-day gardens in California, Florida, and the Southwest; formal gardens of France; Dutch tulip gardens; three centuries of English gardening and their influences on American styles; the beginnings of Naturalism; American gardening from early Colonial times to the city gardens of today. These discussions involve the gardening literature and philosophy of the various eras, the plants and flowers that enjoyed special favor, and the fascinating personalities that played important roles.

And they are all illustrated with more than 100 photographs and drawings, including a Byzantine tree fountain, a medieval may tree, examples of topiary art and hedge sculpture, an early Japanese garden, Tudor garden layouts, a modern German version of the traditional rock garden, summer houses in an Indian garden, a New York terraced garden, winter and summer treatment of orange trees in 17th-century Holland, Renaissance Florentine formal gardens, and many other uncommon gardens, furnishings, and decorative embellishments used through the centuries.

Full of exciting garden ideas as well as a wealth of gardening lore, this history of the long and surprisingly complex evolution of garden types and techniques is certain to please all amateur and professional horticulturists. It is surely a must for the shelves of everyone who has a home garden of his own—or would like to have one.

"No contemporary American could have told the story so well . . . delightful reading," Saturday Review of Lit. "A rare pleasure," Boston Transcript. "An amazing, a superb book . . . a masterpiece of writing," Books.

Unaltered, unabridged republication of original edition. List of full-page illustrations. Total of 104 illustrations. Extensive bibliography. Index. x + 475pp. 5⅜ x 8½.

21105-3 Paperbound **$2.50**

LIFE HISTORIES OF NORTH AMERICAN MARSH BIRDS

by Arthur Cleveland Bent

The all-inclusiveness of Bent's volumes on North American birds has made them classics of our time. Arthur Cleveland Bent was one of America's outstanding ornithologists, and his twenty-volume series on American birds, published under the auspices of the Smithsonian Institution, forms the most comprehensive, most complete, and most-used single source of information in existence. No ornithologist, conservationist, amateur naturalist or birdwatcher should be without a copy.

In this volume, the reader will find an encyclopedic collection of information about fifty-four different kinds of marsh bird (flamingo, ibis, bittern, heron, egret, crane, crake, rail, coot, etc.). Not a group of general descriptions but a collection of detailed, specific observations of individual flocks throughout the country, it describes in readable language and copious detail the nesting habits, plumage, egg form and distribution, food, behavior, field marks, voice, enemies, winter habits, range, courtship procedures, molting information, and migratory habits of every known North American marsh bird.

Completely modern in its approach, the study was made with the full recognition of the difficulties inherent in the observation and interpretation of wild life behavior. For that reason, not only the reports of hundreds of contemporary observers throughout the country were utilized, but also the writings of America's great naturalists of the past — Audubon, Burroughs, William Brewster. The complete textual coverage is supplemented by 98 full-page plates containing 179 black-and-white photographs of nesting sites, eggs, and the young of important species at various stages in their growth, etc.

Unabridged republication of 1st edition. Index. Bibliography. 98 full-page plates. xiv + 392pp. of text. 5⅜ x 8½.   21082-0 Paperbound **$4.50**

LIFE HISTORIES OF NORTH AMERICAN GALLINACEOUS BIRDS

by Arthur Cleveland Bent

The all-inclusiveness of Bent's volumes on North American birds has made them classics of our time. Arthur Cleveland Bent was one of America's outstanding ornithologists, and his twenty-volume series on American birds, published under the auspices of the Smithsonian Institution, forms the most comprehensive, most complete, and most-used single source of information in existence. No ornithologist, conservationist, amateur naturalist or birdwatcher should be without a copy.

In this volume, the reader will find an encyclopedic collection of information about eighty-eight different gallinaceous birds (partridge, quail, grouse, ptarmigan, pheasant, pigeon, dove, etc.). Not a group of general descriptions but a collection of detailed, specific observations of individual flocks throughout the country, it describes in readable language and copious detail the nesting habits, plumage, egg form and distribution, food, behavior, field marks, voice, enemies, winter habits, range, courtship procedures, molting information, and migratory habits of every known North American gallinaceous bird.

Completely modern in its approach, the study was made with the full recognition of the difficulties inherent in the observation and interpretation of wild life behavior. For that reason, not only the reports of hundreds of contemporary observers throughout the country were utilized, but also the writings of America's great naturalists of the past—Audubon, Burroughs, William Brewster. The complete textual coverage is supplemented by 93 full-page plates containing 170 black-and-white photographs of nesting sites, eggs, and the young of important species at various stages in their growth.

Unabridged republication of 1st edition. Index. Bibliography. 93 full-page plates. xiii + 490pp. of text. 5⅜ x 8½.   21028-6 Paperbound **$5.00**

LIFE HISTORIES OF NORTH AMERICAN DIVING BIRDS

by Arthur Cleveland Bent

The all-inclusiveness of Bent's volumes on North American birds has made them classics of our time. Arthur Cleveland Bent was one of America's outstanding ornithologists, and his twenty-volume series on American birds, published under the auspices of the Smithsonian Institution, forms the most comprehensive, most complete, and most-used single source of information in existence. No ornithologist, conservationist, amateur naturalist or birdwatcher should be without a copy.

In this volume, the reader will find an encyclopedic collection of information about thirty-six different diving birds (grebe, loon, auk, murre, puffin, etc.). Not a group of general descriptions but a collection of detailed, specific observations of individual flocks throughout the country, it describes in readable language and copious detail the nesting habits, plumage, egg form, distribution, food, behavior, swimming and diving habits, voice, enemies, winter habits, range, courtship procedures, molting information, and migratory habits of every known North American diving bird.

Completely modern in its approach, the study was made with the full recognition of the difficulties inherent in the observation and interpretation of wild life behavior. For that reason, not only the reports of hundreds of contemporary observers throughout the country were utilized, but also the writings of America's great naturalists of the past — Audubon, Burroughs, William Brewster. The complete textual coverage is supplemented by 12 full-page black-and-white plates showing types of eggs, and 43 plates containing 80 photographs of young at various stages of growth, nesting sites, etc.

Unabridged republication of 1st edition. Index. Bibliography. 55 full-page plates. xiv + 239pp. of text. 5⅜ x 8½.   21091-X Paperbound **$3.00**

LIFE HISTORIES OF NORTH AMERICAN GULLS AND TERNS
by Arthur Cleveland Bent

The all-inclusiveness of Bent's volumes on North American birds has made them classics of our time. Arthur Cleveland Bent was one of America's outstanding ornithologists, and his twenty-volume series on American birds, published under the auspices of the Smithsonian Institution, forms the most comprehensive, most complete, and most-used single source of information in existence. No ornithologist, conservationist, amateur naturalist or birdwatcher should be without a copy.

In this volume the reader will find an encyclopedic collection of information about 50 different gulls and terns. Not a group of general descriptions, but a collection of detailed, specific observations of individual flocks throughout the country, it describes in readable language and copious detail the nesting habits, plumage, egg form and distribution, food, behavior, field marks, voice, enemies, winter habits, range, courtship procedures, molting information, and migratory habits of every known North American gull and tern.

Completely modern in its approach, the study was made with the full recognition of the difficulties inherent in the observation and interpretation of wild life behavior. For that reason, not only the reports of hundreds of contemporary observers throughout the country were utilized, but also the writings of America's great naturalists of the past—Audubon, Burroughs, William Brewster. The complete textual coverage is supplemented by 16 full-page black-and-white plates showing 99 types of eggs, and 77 plates containing 149 photographs of young at various stages of growth, nesting sites, etc.

Unabridged republication of 1st edition. Index. Bibliography. 93 plates. xii + 337pp. of text. 5⅜ x 8½.   21029-4 Paperbound **$4.00**

FRUIT KEY AND TWIG KEY TO TREES AND SHRUBS

FRUIT KEY TO NORTHEASTERN TREES
TWIG KEY TO THE DECIDUOUS WOODY PLANTS OF
EASTERN NORTH AMERICA

by W. M. Harlow

(Professor of Wood Technology, College of Forestry,
State University of New York, Syracuse)

Bound together for the first time in one volume, these handy, accurate, and easily used keys to fruit and twig identification are the only guides of their sort with photographs—over 350 of them, of nearly every twig and fruit described—making them especially valuable to the novice.

The fruit key (dealing with both deciduous trees and evergreens) begins with a concise introduction, explaining simply and lucidly the process of seeding, and identifying the various organs involved: the cones and flowers, and their component parts and variations. Next, the various types of fruits are described—drupe, berry, pome, legume, follicle, capsule, achene, samara, nut—and fruiting habits, followed by a synoptic summary of fruit types.

The introduction to the twig key tells in plain language the process of growth, and its relation to twig morphology through leaf scars, branch scars, buds, etc. For the benefit of the unwary, poison-ivy, poison-oak, and poison-sumac are immediately and fully described.

Identification in both books is easy. There is a pair of alternative descriptions of each aspect of the specimens. Your choice of the fitting one leads you automatically to the next proper pair. At the end of the chain is the name of your specimen and, as a double check, a photograph. More than 120 different fruits and 160 different twigs are distinguished.

This exceptional work, widely used in university courses in botany, biology, forestry, etc., is a valuable tool and instructor to the naturalist, woodsman, or farmer, and to anyone who has wondered about the name of a leafless tree in winter, or been intrigued by an interestingly shaped fruit or seed.

Over 350 photographs, up to 3 times natural size. Bibliography, glossary, index of common and scientific names, in each key. Total of xvii + 126pp. 5⅜ x 8½. Two volumes bound as one.   20511 8 Paperbound 1.35

THE BEHAVIOUR AND SOCIAL LIFE OF HONEYBEES
By C. Ronald Ribbands

This outstanding book offers a definitive survey of all the facets of honeybee life and behavior. The industrious lives of these insects and their highly-efficient, inter-dependent patterns of social organization have made them fascinating objects of study for centuries, and, because they are easily kept and observed, we know more about their ways and habits than we do about any other insect. In this work, Professor Ribbands has brought together the results of experimentation and research from scientific journals and widely-scattered sources from all over the world, and he has presented this accumulated knowledge in interesting, everyday terms that both scientist and layman will appreciate.

Beginning with a basic coverage of physiology, anatomy and sensory equipment—including a full description of the structural differences between worker, queen, and drone, the author also gives a thorough account of the behavior of honeybees in the field. He deals with such questions as how temperature, rain, wind, etc., affect foraging activity and how bees find their way home and back to food areas, covering the details of pollen and nectar gathering, foraging range, mating habits, etc. An important section explains how individuals communicate in various field and hive situations: recruitment to feeding areas (with a discussion of the intricate dances which indicate where these areas are), selection of a home, recognition of companions, the defense of the community, and the like.

There is also an extensive treatment of life within the community, considering the activities of food sharing, wax production, comb building, and brood rearing, the causes of swarming, the queen, her life and relationship with the workers, the evolution of the community and adaptation to a restricted food supply, and many other matters. This is an invaluable book for the beekeeper, natural historian, biologist, entomologist, social scientist, et al. The general reader will find the nontechnical exposition extremely informative and engrossing.

"A 'MUST' for every scientist, experimenter and educator, and a happy and valuable selection for all interested in the honeybee," AMERICAN BEE JOURNAL. "Recommended in the strongest of terms," AMERICAN SCIENTIST. "Erudite, as well as interesting . . . well-documented," NATURAL HISTORY MAGAZINE. "An indispensable reference," J. Hambleton, BEES.

Unabridged, unaltered republication of original edition. 9 photographic plates. 66 figures. Indices. 693-item bibliography. 352pp. 5⅜ x 8½. 21137-1 Paperbound **$3.00**

INSECT LIFE AND NATURAL HISTORY

by J. W. Frost

This very unusual book is a middle-level account of the immense empire of insects and their ways. Although it is used in scores of biology classes throughout the world, it is also an unexcelled browsing or self-study work, full of solid information presented in a most engaging manner.

Professor Frost, who is one of the country's foremost entomologists, presents insects, not as subjects for classification alone, but as living beings, with habits, life cycles, interrlations with other forms of life, and remarkable abilities and deficiencies. Beginning with the biological position of insects throughout the world, he then discusses the origin and distribution of insects, types and classification, the process of metamorphosis and other stages in the life cycle, morphology, insect color, sound, behavior, social life, insects of prey, parasites, insects in relation to plants, subterranean insects, aquatic insects, and similar topics of importance. Very interesting sections are also included on insects as subjects for art, literature, and music, insects as human food, and insects and civilization.

Although Professor Frost's book is thorough and impeccably accurate, it is delightful reading, packed with of startling facts about insects and their ways. You will read about insects that change color at will; insects that feign death; insects that live in hot springs up to 150° moth larvae that construct diving suits for themselves; short circuit beetles that bore through lead electric cables; butterflies whose wing radiation can expose photographic plates; fireflies that flash in unison; insects that live by running after ants and snatching food from their mandibles; insects that are driven by instinct to fly to their death in the snows of Mount Everest, and countless other facts about insects in their relations to each other and to man.

"Distinclty different, possesses an individuality all its own; well-organized, clearly presented, scholarly and pleasing," Journal of Forestry. "Well written, stimulating," Quarterly Review of Biology.

Formerly titled "General Entomology," 1300 item classified bibliography. Appendix with 13 keys enabling you to identify insects. Over 700 illustrations. Index. x + 524pp. 5⅝ x 8⅜.   Paperbound **$4.00**

THE PUMA: MYSTERIOUS AMERICAN CAT

By Stanley P. Young and Edward A. Goldman

The puma (alias "cougar," "mountain lion," "panther," "catamount," etc.) is perhaps the most successful and dangerous of American predators. A deadly hunter of wild animal prey, he is best known to us through the fearsome tales of his attacks on man and livestock; as such he has been the legitimate object of predator control activities from the Yukon to the Straits of Magellan. Yet this magnificent animal's destructiveness and his threat to man's well-being has been somewhat overestimated. This book offers a better understanding of the puma and his mysterious ways. Prepared by two of America's outstanding mammalogists, it is the definitive study of the life-form.

Part I (by S. P. Young) is a comprehensive survey of the puma's history, life habits, and its relationship to man. Making full use of the reports of hundreds of naturalists and woodsmen of the present and past (including Daniel Boone, Lewis and Clark, Darwin, etc.) and the results of numerous government-sponsored field studies spanning more than a quarter-century, the author furnishes exhaustive information on the animal's physical features (coloration, size, weight, strength, etc.), abilities, geographical distribution, breeding habits, enemies, tracks, diseases, food, economic and ecological value, and extensive details on the hunting, trapping, and control of the puma by man. No other book gives you such a complete portrait of this fascinating beast.

The second half of the book (Classification of the Races of the Puma, by the late E. A. Goldman) lists 30 subspecies of Felis concolor, with full descriptions of the distinguishing characteristics of each variety. Also covered are the evolutionary history of the species and its general physical attributes, including photographs and tabulated data on cranial differences, etc. Throughout the book there is a wealth of illustrative material: distribution maps, photos of pumas in their natural habitats, characteristic poses, treed by dogs and hunters, their food, tracks, traps, and the like.

Trappers and hunters, stock raisers, outdoorsmen, natural historians, students, and others will find this an extremely useful and authoritative study of methods of control, identification, and classification, as well as an engrossing account of the puma's life history. Here is also a strong argument—despite the animal's reputation as the leading predator of the New World—for a reasonable conservation approach to this distinctive American species.

Unaltered, unabridged reproduction of original (1946) edition. Foreword. 50-page bibliography of 746 entries. Index. 93 plates, including 165 black-and-white photographic illustrations. 6 figures. 13 tables. xiv + 358pp. 5⅜ x 8⅜. 21184-3 Paperbound **$3.50**